ISO 9001
for Engineers
and Designers

Other ISO Books of Interest

JOHNSON • *ISO 14000 Road Map to Registration*

JOHNSON • *ISO 9000: Meeting the International Standards, Second Edition*

DAVIES • *ISO 9000 Management Systems Manual*

GILLESPIE • *ISO 9000 for the Process Industry*

ISO 9001
for Engineers
and Designers

Stephen J. Schoonmaker

McGraw-Hill

New York   San Francisco   Washington, D.C.   Auckland   Bogotá
Caracas   Lisbon   London   Madrid   Mexico City   Milan
Montreal   New Delhi   San Juan   Singapore
Sydney   Tokyo   Toronto

Library of Congress Cataloging-in-Publication Data

Schoonmaker, Stephen J.
   ISO 9001 for engineers and designers / Stephen J. Schoonmaker.
      p.   cm.
   Includes bibliographical references and index.
   ISBN 0-07-057710-2 (alk. paper)
   1. ISO 9000 Series Standards. I. Title.
  TS156.6.S36 1996
  658.5'62—dc20   96-31723
                                     CIP

McGraw-Hill

A Division of The McGraw·Hill Companies

2 3 4 5 6 7 8 9 10 FGRFGR 9 9 8

ISBN 0-07-057710-2

The sponsoring editor for this book was Harold B. Crawford, the editing supervisor was Patricia V. Amoroso, and the production supervisor was Suzanne W. B. Rapcavage. It was set in Century Schoolbook by Dina John of McGraw-Hill's Professional Book Group composition unit.

Printed and bound by Quebecor

Portions of the Q9001 and Q9004-1 standards are reprinted with the permission of the American Society for Quality Control.

 This book is printed on recycled, acid-free paper containing a minimum of 50% recycled de-inked fiber.

McGraw-Hill books are available at special quantity discounts to use as premiums and sales promotions, or for use in corporate training programs. For more information, please write to the Director of Special Sales, McGraw-Hill, 11 West 19th Street, New York, NY 10011. Or contact your local bookstore.

Contents

Preface

I would like to begin this book by recalling an event that occurred near the beginning of my career. At the time I was employed at a commercial research and development firm near Boston, Massachusetts. After maintaining a set of computer programs for the design of centrifugal compressors, my project manager left the company. As I took over his activities, I needed to prepare a shipment of computer software and documentation to a customer.

I soon discovered, however, that nobody knew where vital user documentation was stored. After a search, I discovered the materials stacked in a cabinet in the company cafeteria! I was amazed. How could this seemingly vital documentation—the only tangible part of the product—have almost been lost? How could the most basic business practices of the firm have been that far out of control?

In the years since that episode, I have found many opportunities to encourage and implement some level of control over technical activities (such as the development of analytical engineering computer software). I also came to the conclusion that the more sophisticated and technologically charged a product or process becomes, the less time and effort is spent trying to understand its development processes and attempting to apply an appropriate level of control over those processes.

In my subsequent experiences, I also found that once computers were involved, few managers wanted to take the time to learn or understand what was going on with a product or project. The computer is the "magic solution," after all. Management's attitude was that the computer was going to fix everything, and they did not have to understand it. By contrast, my opinion was that the more powerful the computers, the more management should learn and understand them. Also, my conclusion was that the more sophisticated computer software became, the more the software should be scrutinized and its development be controlled (or at least understood).

When International Standards Organization (ISO) 9001 burst onto the scene (while I was employed by a large reciprocating compressor design and manufacturing facility in upstate New York), the ISO 9001 auditors were interested in and approved of my newly implemented control over the advanced software activities at that company. It seemed like a vindication of my efforts.

As much as I found myself pleased that my approaches were deemed valuable with respect to the ISO 9001 audit, I also found myself concerned that I had introduced a layer of artificial procedures that merely gave the appearance of proper management. As I thought back to the episode of finding the documentation in the cafeteria, I realized that there may in fact have been nothing wrong in that situation. That particular research and development firm existed to develop advanced new technology, not to document every procedure it ever used (including where to find the documentation). In that case, it might have been a waste of the firm's time and energy to know where those documents were located.

Perhaps gaining control over activities such as design and engineering was not as simple as I had first thought. For instance, if large amounts of effort were spent documenting every procedure or process in a company, there could be an excessive emphasis on control. There may be resistance to changing or eliminating their efficient processes. If excessive procedures were applied to technology development, the situation could have disastrous consequences for a company. A company could fall behind its competition if it wasted effort on the control of functions, and might lose sight of the importance of technological development altogether.

Considering these challenges I have discovered with respect to engineering and its management, I feel compelled to make a few cautionary statements. First, ISO 9001 seeks to document and apply control to all kinds of processes; therefore, those involved with ISO 9001 need to be aware of the importance of balancing the need to control and document and the need to let new ideas and processes evolve. Second, ISO 9001 should not always be considered a problem for organizations to resolve; ISO 9001 can, in fact, be a vehicle for helping organizations understand their own technologically charged processes and implement a balanced level of control over them. Third, I cannot guarantee success with an ISO 9001 registration process by following the procedures and philosophy presented in this book. The procedures, and the philosophy they are based on, represent my best effort to show typical and instructive examples. Considering the vast variety of products, markets, and companies all over the world involved with ISO 9001, my ideas cannot be expected to work in all situations.

Just as the need for control over development processes seemed a simple matter to me at first, the reader (particularly the designer or engineer of advanced technology) should not consider ISO 9001 necessarily a simple (and possibly annoying) matter. He or she should be careful not to think of it as just another management consultant's fantasy where government agencies require companies to get "quality." The situation may not be that simple. ISO 9001 can provide an opportunity to make management sit up and listen to the struggles of the designer or engineer. By using this book, designers and engineers are hopefully going to be able to understand what is involved with ISO 9001 and then take advantage of its requirements. Most designers and engineers know of areas in their development processes that could use some updating or fixing. Why not use an upcoming ISO 9001 audit as the excuse for correcting the situation?

Finally, although I am not a quality professional, I am an engineer that has been actively involved with the preparation of procedures for meeting the ISO 9001 requirements. My practical experience lies in the development, implementation, and maintenance of computer programs for the energy and aerospace markets. I have developed a few successful algorithms for the prediction of the behavior of this type of equipment; I have trained and assisted many engineers and designers in the use of advanced software tools for analysis and design. Also, I have successfully helped in making ISO 9001 registrations a success. But, I must point out that my presentation is naturally inclined toward the mechanical engineering field and my particular experiences.

I would be remiss if I did not make some acknowledgments. I would like to thank my publisher for giving me this opportunity to publish my thoughts and experiences. I must also thank co-workers at the Northern Research and Engineering Corporation (NREC) and Dresser-Rand Engine Process Compressor Division (D-R EPCD). I would specifically mention Jan Tauc and Mel Platt at NREC, and Don Draper and Robert Bennitt at D-R EPCD.

Of course, I have no idea why the Lord of the Universe has allowed me the ability to write a book such as this and to be blessed with a supportive family, but God's knowledge passes all human understanding. I just hope I have used my talents wisely and may be helping others in their careers and struggles in our modern world. Thank you to my wife, Sharon, and the best kids I know—Christina, Michael, Jennifer, and Melissa.

Stephen J. Schoonmaker

ISO 9001
for Engineers
and Designers

Introduction

This book presents a collection of information and examples about the international quality standard referred to as ISO 9001. This presentation is intended for the benefit of technological specialists such as designers and engineers. This presentation may also be of interest to executives, managers in technical environments, and quality assurance specialists. Although there are many sources of information on ISO 9001, this book strives to give a unique technological perspective.

Why ISO 9001?

Many people refer to ISO 9000 and ISO 9001 interchangeably. This is particularly true for designers and engineers since the advent of the ISO 9000 phenomenon for these professionals has generally meant their companies have been seeking registration to the ISO 9001 standard. However, one should understand that ISO 9000 is not even a standard, and companies are not audited for ISO 9000 compliance. Instead, ISO 9001 is a specific standard in the ISO 9000 family of quality standards, and it is ISO 9001 which is the basis of these company's audits. Furthermore, ISO 9001 is the only standard in the family that applies to activities called design and development. Therefore, it is obvious that this book (which is intended for designers and engineers) is focused on the ISO 9001 standard.

Although ISO 9001 is the focus of this book, other documents in the ISO 9000 family are also discussed. In particular, ISO 9000 itself is a guideline document that provides some background and fundamental information. This information is presented as valuable for providing an understanding of where ISO 9001 is coming from. ISO 9004 is a useful source of introductory information, and an entire chapter

(Chapter 9) discusses ISO 9004 in the context of learning about quality systems.

Practical Application

In order to have the most efficient use of the information in this book for designers and engineers, this book moves as quickly as possible into a section on the practical applications and ramifications of ISO 9001. The discussion in that section of the book and its sample procedures are based on the framework of an "ideal" ISO 9001 process. This ideal process is a product design and development management model based directly on the ISO 9001 standard. This ideal process is not expected to be implemented into an actual company or organization, however. Instead, this idealized process is intended to show all the concepts and practical implications of the ISO 9001 standard. Hopefully this is the most efficient means of presenting this material to the intended audience.

This ideal ISO 9001 design process and its resultant sample procedures and documentation are covered in Chapters 3 through 7. Readers intending to study information directly relating to designers and engineers should refer to these chapters (preferably while referencing a copy of the ISO 9001 standard).

Quality Management

Chapters 3 through 7, which present the ideal ISO 9001 process, are intended to help an organization's designers and engineers understand, apply, and quickly take advantage of ISO 9001. However, these chapters attempt to minimize the details of the standard itself. The subsequent chapters (8 through 10), however, present a more thorough discussion of quality management, quality systems, and the ISO 9001 standard. These later chapters may be of more interest to managers of designers and engineers as opposed to practicing designers and engineers. However, some designers and engineers may wish to learn more about the standards than just surviving audits, and these later chapters should prove useful in this regard as well. In addition, Appendix B presents thoughts about how design and engineering functions deal with or even conflict with the ISO 9001.

2

Background

Before presenting detailed information about the design and development requirements of the ISO 9001 standard, we may wish to consider some background for this standard and the group of standards it belongs to under the heading of ISO 9000. This chapter attempts to present some of this background information; however, the reader should note that ISO 9000 is a changing and evolving set of standards. A primary source of information on ISO 9000 standards should be sought whenever possible.[1]

The ISO 9000 Phenomenon

The acceptance and growth of ISO 9000 can only be characterized as phenomenal. It has taken a significant portion of the worldwide industrial complex by storm. ISO 9000 has attracted significant attention among business managers since its appearance in 1987. Regardless of where it has come from, or why it has appeared, it has already affected thousands of organizations and companies, as well as dozens of countries and governments.[2]

Globalization

The phenomenal impact of ISO 9000 may be attributed to a number of factors. Perhaps the most important is the globalization of business and economics. As national boundaries have become more relaxed and as communications technologies have rapidly advanced, global business arrangements have been created that are more tightly integrated internationally than in the past. Instead of setting up subsidiaries as virtually independent companies in other parts of the world, operations in distant parts of the world can now be made fully functional components of a truly worldwide organization.

Although these worldwide companies have probably not demanded something like ISO 9000 in order to qualify their in-house operations across the world, these companies have probably paved the way for the globalization of business in general. Now companies may be dealing more often with operations beyond their national borders, and are · rushing to develop their markets on a global scale. As part of this globalization process, companies have apparently recognized a need to ensure basic capability and competence among operations that are separated by vast geographic distances. Assuming these companies are looking for the basic ability to meet customer demands or requirements, the ISO 9000 quality system standards seem to fit the need for qualification of companies. However, the degree to which companies are deciding to not deal with other companies just because they do not have ISO 9000 registration is probably not conclusively known.

Quality principles

Another major factor in the ISO 9000 phenomenon is the propagation of quality principles or ideas. As opposed to the older idea that a company inspects its product after it is manufactured to see that it meets specifications, businesses have turned their emphasis to ensuring that the product is correct in the first place (before any inspection). This shift in emphasis is generally based on the idea that one must develop a capable or quality-driven organization. Such an organization is supposed to produce correct results the first time. Cost savings are supposed to be realized by not instituting the quality control effort after it is too late, and customers are supposed to be more likely satisfied with the final product.

ISO 9000 seems to completely support this notion of having a quality-capable organization. ISO 9000 attempts to specify the minimum levels of quality management in organizations. Then ISO 9000 requires that the organizations maintain these capabilities and make a commitment to their continuous improvement. ISO 9000 also requires third-party or independent audits for verification of these capabilities. ISO 9000 seems to be based on the idea of inspecting the processes and personnel producing the product, instead of the product itself.

ISO 9000 has been perceived as a means of demonstrating that an organization has quality management, and it is supposed to demonstrate that the organization is committed to quality or customer satisfaction ideals. This has probably been a contributor to its wide-scale acceptance. To what degree ISO 9000 actually indicates that an organization is meeting those customer satisfaction ideals is probably not known with certainty.

Economic and political goals

For better or worse, companies, governments, and independent agencies such as standards bodies and professional societies have also perceived ISO 9000 as a means of pursuing their own goals in the marketplace. For some companies, ISO 9001 has been seen as a method of demonstrating a competitive advantage. If one company is ISO 9001 registered and others in the market are not, the registered company may be perceived as more likely to produce quality designs and products. Of course, not having ISO 9001 registration does not mean that a company's products are of poor quality, but companies may not want to risk adversely affecting their market position by not getting registered. This corporate peer pressure may have some bearing on the wide acceptance of ISO 9000 as well.

From its first significant growth in the United States, ISO 9000 has been perceived as a means of pursuing governmental objectives. One commonly heard example is that ISO 9000 is a means for European protectionism. According to this view, the European political and economic system decided to require ISO 9000 standards registration for businesses importing to Europe as a veiled attempt to restrict imports. ISO 9000 registration is seen as a completely new import requirement that started in 1989, and the addition of this requirement came at the time when few, if any, companies in the United States had even heard of ISO 9000.[3]

The view that ISO 9000 has been a vehicle for pushing European political and economic goals seems to fit the author's experience with the ISO 9001 standard in the late 1980s. As a seemingly new requirement for selling equipment to a major energy concern in the Netherlands, the author's employer at the time was expected to be registered to an ISO 9000 standard. Considering this American manufacturer's success in exporting to and integrating with the European market since the early 1900s, this requirement seemed to have no other purpose than to restrict access to the European market and protect local manufacturers who were competing with the American firm. The quality and safety of the American equipment had already been established for decades by the marketplace, so what good would be done by adding a layer of non-value-added government requirements?

The final apparent contributor to the growth of the ISO 9000 standards is U.S. government streamlining. As many parts of the federal government have quality systems requirements [Department of Defense, National Aeronautics and Space Administration (NASA),

Federal Aviation Administration (FAA), Nuclear Regulatory Commission 10CFR50, etc.], it seems that a consolidation and rationalization has been envisioned by utilizing the ISO 9000 standards. Indeed, ISO 9000 has gained acceptance by at least the Department of Defense, and older quality system documentation can apparently be replaced by ISO 9000 standards.

What Is ISO 9000?

As mentioned throughout this book, ISO 9000 is not a standard. It is a generic term applied to a series of standards and accompanying guideline documents. Figure 2.1 shows these standards and guidelines. Note that there is an ISO 9000 document itself entitled *Quality Management and Quality Assurance Standards—Guidelines for Selection and Use* (see Bibliography). This document introduces and explains the philosophy of the ISO 9000 series.

ISO is the French-language acronym for the International Organization for Standardization. It is headquartered in Geneva, Switzerland. With respect to quality systems, the American Society for Quality Control (ASQC) is the member body to the ISO from the United States. Thus, the ASQC publishes the ISO 9000 series or family of documentation, and the ISO 9001 standard is published as an American National Standard (or an ANSI standard) under the designation ANSI/ASQC Q9001-1994.

The ISO 9000 series is conceived, developed, prepared, and improved by a technical committee (TC) within ISO. This particular committee is called ISO/TC 176. This committee has participants from 41 countries, and 22 other countries have an observer status. This committee has also worked with the International Electrotechnical Committee (IEC/TC 56).[4]

The TC 176 has three subcommittees. The first subcommittee (SC 1) deals with terminology; this subcommittee is run by Association Francaise de Normalisation (AFNOR) (the French standards organization) and has working groups convened by the British Standards Institution (BSI) and the ANSI. The second subcommittee (SC 2) develops quality systems; it is run by BSI and has working groups convened by experts from the BSI, ANSI, a Canadian Standards Organization, and Deutsches Institut für Normung (DIN) (Germany). The third subcommittee (SC 3) is responsible for tools for implementation; this group is run by a standards organization from the Netherlands, and it has working groups run by experts from BSI, ANSI, and DIN.[5]

As Figure 2.1 shows, the actual standards within the ISO 9000 series are ISO 9001, ISO 9002, and ISO 9003. Clearly ISO 9001 is the

ISO 9000	Quality Management and Quality Assurance Standards - Guidelines for Selection and Use
ISO 9001	Quality Systems - Model for Quality Assurance in Design, Development, Production, Installation, and Servicing (Standard)
ISO 9002	Quality Systems - Model for Quality Assurance in Production, Installation, and Servicing (Standard)
ISO 9003	Quality Systems - Model for Quality Assurance in Final Inspection and Test (Standard)
ISO 9004	Quality Management and Quality System Elements - Guidelines
ISO 9000-2	Quality management and quality assurance standards - Part 2: Generic guidelines for the application of ISO 9001, ISO 9002, and ISO 9003
ISO 9000-3	Quality management and quality assurance standards - Part 3: Guidelines for the application of ISO 9001 to the development, supply and maintenance of software
ISO 9004-2	Quality management and quality system elements - Part 2: Guidelines for services
ISO 9004-3	Quality management and quality system elements - Part 3: Guidelines for processed materials
ISO 9004-4	Quality management and quality system elements - Part 4: Guidelines for quality improvement

Figure 2.1   ISO 9000 series of documents.

most comprehensive of the standards. It covers a wide range of activities related to product quality including design, manufacturing, installation, and service. As stated a number of times in this book, it is ISO 9001 that is of critical concern to technical professionals such as designers and engineers, and this standard is therefore the central focus of this book.

The current version of the ISO 9000 series of standards is 1994. Its approval date is listed as August 1, 1994.[6] Furthermore, the TC is supposed to revise the series every 5 years; therefore, the next edition should be in force around the year 2000.

It is important to realize that compliance with a standard in ISO 9000 involves an audit. To obtain the official registration for ISO 9001, for example, an organization has an accredited, independent auditor or registrar perform an on-site inspection or audit. These audits are generally repeated at 6-month intervals. In the initial audit and at intervals such as every three years, a comprehensive audit is conducted throughout diverse departments within the organization. At the 6-month intervals, specific smaller areas within the organization may be investigated.[7]

ISO 9000 and Various Business Sectors

Various public and private business sectors appear to have some relationship with the ISO 9000 series of standards. This section presents a brief description of some of these relationships.

Pressure vessels, elevators, fasteners, nuclear, and offshore equipment sectors

The American Society of Mechanical Engineers (ASME) has been an important body for standards since the boiler test code was created in 1884. For over 75 years, the ASME has also been accrediting quality systems of manufacturers for basic businesses such as boilers and pressure vessels, elevators, fasteners, nuclear power plants, and offshore oil and gas equipment. This accreditation involves the ASME's published codes and standards, and it culminates in the right to use an ASME stamp on components such as pressure vessels. Parts of this process involve on-site inspection and auditing as well as the accreditation of quality management activities.

One can see the impact and influence of the ISO 9000 standards, therefore, by the fact that the ASME chose to become an ISO 9000 registrar. The ASME can now be called on to conduct a supplier registration for ISO 9000 standards. It was accredited by the Registrar Accreditation Board (RAB) and the Dutch Council for Certification (RvC).[8] The ASME ISO 9000 registration program offers registration of suppliers of mechanical equipment and related materials in sectors addressed by ASME codes and standards (such as the ASME Pressure Vessel Code).

Since documents such as the ASME Pressure Vessel Code are specialized for the calculation and evaluation of specific mechanical components (such as calculations of stresses and strengths of components to assure their integrity), there is no direct conflict between these types of codes and standards and ISO 9000. ISO 9000 clearly does not contain stress calculation types of information.

However, for codes and standards that contain information about quality systems (which might appear in a code or standard with respect to handling of records and procedures in general), ISO 9000 may be connected with such codes and standards already. The ASME Nuclear Quality Assurance (NQA) document in particular may develop a close connection to ISO 9000. This situation should be monitored by any organization connected with the nuclear business sector.[9,10]

Defense sector

As with the sectors involved with nuclear power, ISO 9000 is developing a relationship with the defense sector. U.S. government streamlining programs have been developing to replace military standards with commercial standards. MIL standards for quality assurance can now be covered by ISO 9000 or Q9000 standards. This has affected standards such as MIL-Q-9858A and MIL-STDs 973 and 498. Although specific organizations should check with their status, it is clear that the ISO 9000 standard is likely to be a key component in future military contracts. The relationship of DODI 5000.2 to ISO 9000 should also be monitored for firms involved with the defense sector.

Automotive sector

The automotive sector in the United States clearly now has a relationship with the ISO 9000 standards. In 1994, a quality-related document called QS-9000 was developed by a task force for the three main U.S. automotive companies (Chrysler, Ford, and General Motors). This document defines their basic quality system requirements.

QS-9000 uses ISO 9001 as its foundation. One section of QS-9000 includes the exact text of ISO 9001 with added automotive requirements. The automotive companies are expected to use third-party registration as well. Obviously, then, designers and engineers involved in the automotive sector need to be concerned with the ISO 9001 standard and keep abreast of its developments.

Conclusion

This chapter has presented some background information about the ISO 9000 system of standards. This has included its impact on the world market, the advent and maintenance of the standards, and some of the relationships between various industrial sectors and ISO 9000. The growth of these standards has been remarkable. Designers and engineers should attempt to keep informed about these standards since their future revisions may produce even greater impact on design, engineering, and global markets.

Subsequent Chapters 3 through 7 present a framework for implementing ISO 9001 requirements in the design and engineering environment. Chapters 8, 9, and 10 of this book look at the ISO 9000 philosophy, the ISO 9004 quality system guidelines, and the ISO 9001 standard in particular.

References

1. Two official sources of information are the following:

    American Society for Quality Control (ASQC)
    611 East Wisconsin Avenue
    Milwaukee, Wisconsin 53202

    ISO Central Secretariat
    1, rue de Varembe
    Case postale 56
    CH-1221
    Geneva 20, Switzerland

2. According to the ASQC's "journal of record," *On Q* (May 1995, p. 8), more than 70,000 organizations had earned registration to the ISO 9000 standards by mid-1994.
3. According to the ASQC's *On Q* (November 1992, p. 4), 48 percent of U.S. firms still had not heard of ISO 9000 at that time (5 years after the benchmark 1987 version).
4. Interview with Reg Shaughnessy and Peter Ford published in *ISO 9000 News,* Vol. 1, No. 3, May 1992, p. 3.
5. Interview with Reg Shaughnessy and Peter Ford published in *ISO 9000 News,* Vol. 1, No. 3, May 1992, pp. 2–3.
6. ASQC, *ISO 9001 (ANSI/ASQC Q9001-1994) Quality Systems—Model for Quality Assurance in Design, Development, Production, Installation, and Servicing,* ASQC, Milwaukee, Wisconsin, 1994, p. i.
7. ASQC, "9000 Standards?," *Quality Progress,* Vol. 29, No. 1, January 1996, p. 25.
8. ASME, *ASME NEWS,* Vol. 13, No. 10, February 1994, p. 1.
9. Contact Council on Codes and Standards, ASME, New York, New York 10017.
10. According to June Ling, *ASME NEWS,* Vol. 12, No. 11, March 1993, p. 1, "the Council on Codes and Standards is continuing its work with the U.S. government regarding international trade of regulated products manufactured to ASME codes, standards, and related conformity assessment programs." It is stated on p. 1 in the same issue that "While ISO 9000 registration alone will be insufficient for regulated products, it will most likely be a basic requirement."

3

The Ideal ISO 9001
Design Process

Introduction

This chapter presents a process which is intended to meet the design
and development requirements of ISO 9001[1] as interpreted by the
author. This process should be considered an ideal since it is based
directly on ISO 9001. In most cases, design or engineering departments already have established their own processes and practices,
and the concepts presented here would need to be incorporated or
adapted into that existing process. Of course, if no formal process
already exists, the ideal process could be attempted in the form presented here. Finally, later chapters of this book show that there are
many other requirements in ISO 9001 besides those concerned with
design and development process requirements. The reader should
keep in mind that these other requirements must still be met in order
to be successful with ISO 9001.

Overview

Figure 3.1 shows an overall view of the proposed ideal process. This
most basic view can and should be compared to the ISO 9001 standard paragraph 4.4 on Design Control. This paragraph of the ISO
9001 standard is by far the most crucial area of the standard for the
design and engineering department of an organization. Indeed, this
paragraph must be completely understood by those responsible for
design and engineering. The Design Input, Design Output, and
Design Validation stages shown in Figure 3.1 correspond directly with
subparagraphs of the ISO 9001 standard (see Figure 3.2). It should be

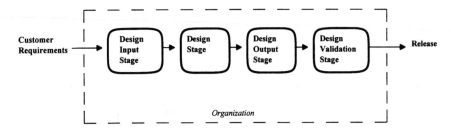

Figure 3.1   The ideal ISO 9001 Design and Development Process.

noted, however, that the Design stage (where the actual design work is performed) is not actually a subparagraph; it is simply assumed to exist in the process.

Figure 3.1 shows a dashed line around the activities in the ideal process. The dashed line represents the boundaries of the organization performing these activities. Beyond this boundary lies the external customer. Indeed, this figure is intended to be interpreted from the perspective of the external customer. Any real design and development process is much more complex than the one shown, but to the external customer (or perhaps an auditor) this is a reasonable approximation of the process.

One can consider the external customer to be a specific, single customer which has contracted with the organization for a single project. Or the external customer might be an entire market that is researched, focused, and crystallized into a perceived demand for a particular new product. In either case, the process is shown to start with the customer requirements. In the first case, the customer

Ideal ISO 9001 Process Activities	ISO 9001 Paragraph
Design Input	4.4.4
Design Output	4.4.5
Design Validation	4.4.8
Design Verification	4.4.7
Design Review	4.4.6

Figure 3.2   Ideal ISO 9001 Process references to ISO 9001.

requirements could be what is found in a contractual agreement. In the second case, the customer requirements could be a list of perceived customer needs based on market research. As shown in Figure 3.1, the customer requirements enter into the organization and go directly to the Design Input Stage. In some cases, this entry may not be directly into the design and engineering department of the organization but rather into the marketing or customer coordination department. However, in this idealized process, it is assumed that the customer requirements flow directly into the design and engineering department where the design is to be created to meet those requirements.

Continuing to look at the process in Figure 3.1 from the perspective of the customer, the end result of the process is the "release." This represents handing over the product to the customer or market. This is the point at which the quality of the product is determined. In the author's mind, the purpose of ISO 9001 (and quality systems in general) is to ensure that what is released meets the customer's requirements. Thus, the output of the process in Figure 3.1 (release) is to satisfy the input of the process (customer requirements).

An engineer viewing Figure 3.1 may be tempted to view it as a control diagram or flow chart. One might further want to assume that a feedback mechanism can be used to have the desired process output meet the desired process input. This would mean assuming that the product is released and then it is tested by the customer. Depending on the product's success or failure with the customer, it would be assumed that the design can be changed until the desired results are achieved.

This feedback mechanism may be fine for a course on control systems, but this is a poor strategy for success in the marketplace. Instead of letting the customer test the product, we should attempt to devise a system that minimizes the amount of product deficiency (or worse yet lack of safety) that is to be discovered by the external customer. In other words, we must *not* draw a line from the output of the process in Figure 3.1 to the input (outside the dashed line) since it implies that the customer is the first to determine if the design is acceptable. Also, the idea that quality means meeting customer requirements alone is not sufficient; quality should mean meeting the customer's requirements the first time the customer interacts with the product. The process inside the dashed line of Figure 3.1 should strive to make this result a reality. Of course, there are other activities within the organization that need to satisfy the customer (such as production or manufacturing), but clearly all other efforts are doomed if the design is not basically capable of attaining customer satisfaction.

Beginning with the stage called Design Validation, then, the activities within the stages of the ideal process are expanded upon for the ideal process being presented. The reader should keep in mind that the goal of these activities is going to be assuring that the design is acceptable within the boundaries of the organization.

Design Validation Stage

Referring to Figure 3.3, it is seen that the Design Validation Stage is expanded to show validation activities followed by a check or decision point. This Design Validation Stage would not be visible to an external customer representative since it is entirely within the organizational boundary. Although some products may use test marketing or require customer witness tests, there could be an unacceptable level of exposure (to methods, materials, or technology, etc.) in this technique. Whether or not a test market should be involved in the Design Validation Stage is a decision each organization must make based on its own analysis of the design and the marketplace, while keeping in mind the goal of customer satisfaction on the first attempt.

As shown in Figure 3.3, the Design Validation Stage contains a check to see if the product meets (or exceeds) the customer requirements. This is the activity which the earlier discussion stated should not be done by the customer externally. Instead this activity is brought within the organization. Also, the reader could now consider this a feedback loop of sorts. If the organization feels that the product is not going to meet the customer requirements, then the organization should reenter the Design Input Stage or the Design Stage as needed. If the organization feels that the product is going to meet the requirements, then the product can be released for final production and then enter the installation and service phase of its life cycle.

A review of paragraph 4.4 of ISO 9001 reveals that two activities are not yet included in the ideal ISO 9001 process. These activities are not considered separate large stages by the author, but they are instead incorporated into the four larger stages. These activities are Design Review and Design Verification. Design Verification is incorporated into the process in the Design Stage as shown in Figure 3.4, and Design Review is incorporated into the process in the Design Output Stage as shown in Figure 3.5.

Design Stage—Design Verification

As can be seen in Figure 3.4, the Design Verification activity is contained within the Design Stage. The product design activity (determining a product's proposed size, shape, behavior, materials, etc.)

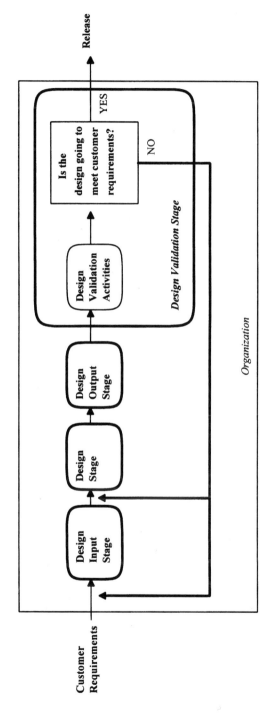

Figure 3.3 The Design Validation Stage.

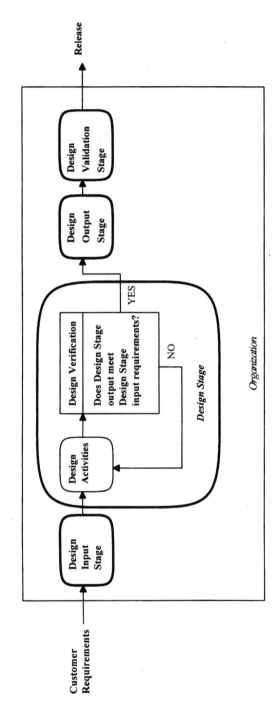

Figure 3.4 The Design Stage.

16

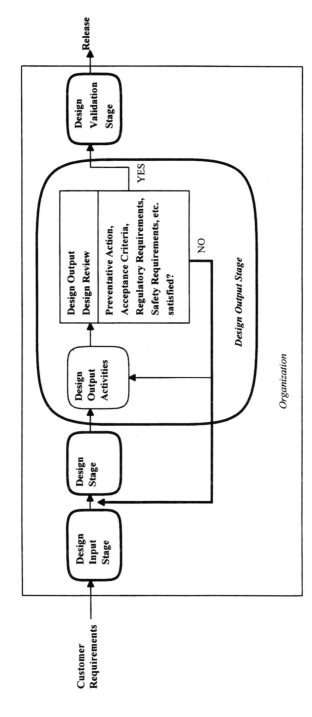

Figure 3.5 The Design Output Stage.

17

obviously is also found in the Design Stage. As paragraph 4.4.7 of ISO 9001 indicates, the Design Verification activity is to ensure that the output from the Design Stage meets the requirements of the Design Stage input. This input, in turn, is the output from the Design Input Stage.

One should note that the Design Verification activity is even more removed from the customer perspective. This activity is not only within the organization's boundary, but it is also likely to be within the innermost circles of the design department of the organization. As Figure 3.4 shows, this activity is like an "inner loop" check on the design, while the Design Validation activity is like an "outer loop" check on the final product. To give some indication of the nature of this activity, paragraph 4.4.7 of ISO 9001 mentions technically oriented activities such as using alternative calculations for the design parameters, using tests and demonstrations, and using comparisons with existing successful designs. As discussed below, the Design Verification activity must also at least include a meeting called a design review.

The Design Verification activities are assumed to be performed with respect to the specifications outlined from the Design Input Stage (which may only be indirectly linked to the customer requirements). On the other hand, Design Validation is to be performed directly with the external customer requirements. Although Design Input and customer requirements should be related, this is a helpful distinction in understanding the basic difference between Design Verification and Design Validation.

Design Reviews

The next activity to be folded into the ideal process is Design Review. Although not explicitly stated as such in the ISO 9001 standard, the Design Review activity is clearly a meeting of some sort. At this meeting (or series of meetings), participants from appropriate departments critique the design. This meeting is documented (with minutes, for instance), and the documentation is expected to be controlled and made available to auditors to show that the review was actually held.

The Design Review activity is perhaps the most confusing concept within the design control section of the current version of the ISO 9001 standard. This is because it is listed as a separate activity in paragraph 4.4.6 to be performed at "appropriate" times, but then it is also mentioned as a requirement in conjunction with the Design Verification paragraph. In the author's mind, this simply means that the intent of the Design Verification activity (to determine if Design Stage output meets Design Stage input) can be met by a design review activity, and that at least this form of design verification must be planned, conducted, and documented.

Although conducting a design review meeting for the Design Verification activity could be viewed as the only design review meeting strictly required in ISO 9001, an ideal process description is being presented here, not a minimal process. The author feels that a specific Design Review activity (in addition to Design Verification) should be incorporated into the Design Output Stage. As Figure 3.5 shows, this activity is completely contained within the Design Output Stage, and it follows the completion of the design output activity (as explained shortly, design output entails such activities as the preparation of the complete design documentation—drawings and bills, etc.).

This Design Review activity shown in Figure 3.5 is the last chance before the Design Validation Stage to review the design. Recall that once the Design Validation Stage is reached, the design is being evaluated in the end-user environment (and perhaps in view of the customer). It is likely to be easier to request a change to the design within the Design Output Stage than to wait for the Design Validation Stage. The design and engineering department of the organization may also have less control over decisions within the Design Validation Stage. For at least these reasons, it is important for a Design Review activity to be held at the end of the Design Output Stage.

Design Output Stage

To this point, the Design Stage, the Design Validation Stage, the Design Verification activity (part of the Design Stage), and the Design Review activity (part of the Design Output Stage) of the ideal process have been briefly introduced. The next stage to be introduced is the Design Output Stage. As mentioned earlier, the Design Output Stage is where the design and engineering department of the organization prepares the documentation package for a product's design. The package generally includes at least a set of two-dimensional drawings and bills of material which document the design and engineering departments intent. The package could also include material specifications, three-dimensional models, specific vendor information, technical manuals, installation instructions, product safety directive information, manufacturing instructions, etc. Although it is going to vary widely for different organizations and markets, this stage may require as much time and effort as the Design Stage.

Paragraph 4.4.5 of ISO 9001 covers Design Output, and it makes certain demands on this stage's activities. These demands include checking that the Design Output meets the Design Input requirements. This is why the Design Review activity was added to the Design Output Stage earlier. Another demand for the Design Output Stage is containing and referencing acceptance criteria. These are specific requirements demanded by the customer or demanded by regulatory agencies, for instance. Meeting these requirements can not be checked in the Design Review activity if they are not clearly stated. Therefore, a part of the Design Output Stage should be the preparation of an acceptance criteria report or document.

Finally, ISO 9001 demands that the Design Output Stage identify safety-related issues in all aspects of the product's life cycle. Meeting these requirements again can not be checked in the Design Review activity if they are not clearly stated. Therefore, a part of the Design Output Stage should be the preparation of a safety review report or document. Then, the standard design output documentation (drawings, bills, etc.), the acceptance criteria document, and the safety review document can be used in the Design Review at the end of the Design Output Stage. As mentioned earlier, this Design Review can then be held and documented. If there are problems discovered from this review meeting, corrective action can be taken (returning to Design Input Stage or Design Stage, for instance).

Design Input Stage

The final stage to be presented in the ideal ISO 9001 product development process is the Design Input Stage. The task of the Design Input Stage is to prepare the requirements of the design from the internal perspective of the organization. It is assumed that the organization knows what needs to be designed based on the external customer requirements. However, there is often an interpretation process that needs to be performed based on those customer requirements. This is the task of the Design Input Stage. The Design Input Stage is to refine those customer requirements into design specifications and create the documentation needed for the later stages. The output from the Design Input Stage can be referred to as a design specification (document or package). The Design Input Stage activity may also be based on a specific contract for a given project.

The Design Input Stage (through the design specification) is likely to generate specific numerical limits for the design and/or specific performance targets. This is a critical process which can make or break the product's design. The design specification must meet the customer requirements, but some organizations may decide to exceed those expectations. Other organizations may decide to reject the customer requirements as too difficult. In any case, it is easiest to alter the design specification at this stage, so the Design Input Stage must be given as much attention as possible and the thinking of those involved in the activity must be proactive.

In addition to specifying the obvious physical design specifications based on the customer requirements, the Design Input Stage is required by ISO 9001 to include other requirements as well. These requirements include issues beyond the customer relationship such as statutory and regulatory requirements. Thus the Design Input Stage involves preparing a complete design specification document that can be used throughout the rest of the process (since the later activities need to know such requirements as acceptance criteria and safety regulations). Figure 3.6 shows the activities within the Design Input Stage.

Figure 3.6 also shows an Input Review activity at the end of the Design Input Stage. ISO 9001 requires that the design specification be reviewed for adequacy. The specification is considered inadequate if it is ambiguous or incomplete. The purpose of the Input Review activity is to check for this adequacy. As discussed later, there may be other review or "quality planning" activities employed at this stage (depending on the nature of the design projects typically engaged in by the organization).

Complete Ideal Process

A presentation of the ideal process is now complete and Figure 3.7 shows the complete process. Also, Figure 3.8 shows some typical information or documentation that is flowing from one stage to the next. One should note that this ideal process is assumed to be followed for every development project within the organization. That is, every product design is expected to go through each of the activities presented (at least based on the author's interpretation of the standard). Of course, some organizations may undertake projects of differing levels of complexity. Hopefully the presented ideal process sheds sufficient light on the demands of ISO 9001 to permit a satisfactory process to be realized by such organizations.

In conclusion, an ideal product development process has been presented. This consists of four major stages. These stages contained various activities. The development of the process using the process flow diagram is related to the input, activity, and output model found in the quality system guideline document (the reader can refer to Figure 1 of ANSI/ASQC Q9000-1-1994).

Organizational Demands

The ideal product development process thus presented makes use of most parts of the ISO 9001 standard, paragraph 4.4 on Design

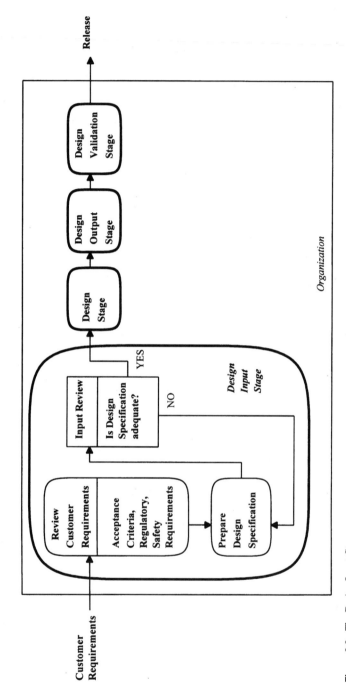

Figure 3.6 The Design Input Stage.

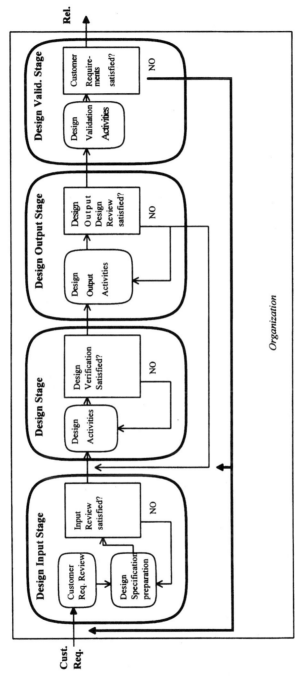

Figure 3.7 Ideal ISO 9001 Process expanded view.

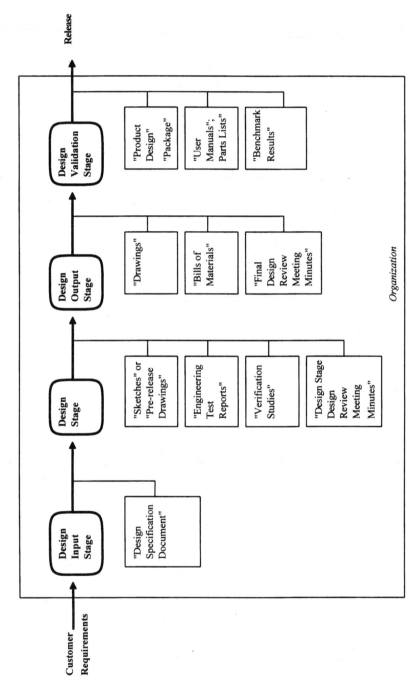

Figure 3.8 Ideal ISO 9001 Process input and output.

Control (see Figure 3.2). However, the ideal process did not touch on the concepts in the remaining parts of paragraph 4.4 (which still impact on the design and engineering effort of an organization). These organizational demands are Documented Procedures, Design Plans, Adequate Resources, Interfaces, and Revision Control (Figure 3.9). Although they do not always pertain to the process followed by every product designed, they are important just the same for success with ISO 9001 certification.

These concepts are viewed as pertaining to the workings of the design and engineering department in general. There should be checks at regular intervals to make sure that each concept is being adequately addressed. At these intervals, documented evidence of the review (called a process review, for instance) would have to be created (such as meeting minutes or a report). If an organization does large-scale design projects that last for long periods of time or projects that require a new design and engineering team or process be formed for each project, then these concepts would be reviewed at the beginning of each product development project. In this case, this process review must be added to the ISO 9001 ideal process (between the Design Input Stage and the Design Stage, for instance). In an organization that does many product development projects in a short amount of time, this procedure may not be practical; in that case the process review could be done at regular intervals instead (such as monthly or yearly).

Organizational Demand	ISO 9001 Paragraph
Documented Procedures	4.4.1 General
Design Plans	4.4.2 Design and Development Planning
Adequate Resources	4.4.2 Design and Development Planning
Interfaces	4.4.3 Organizational and Technical Interfaces
Revision Control	4.4.9 Design Changes

Figure 3.9 Organizational demands references to ISO 9001.

Documented procedures

Paragraph 4.4.1 of the ISO 9001 is the first general concept in Design Control, and is entitled General. It covers the concept of Documented Procedures. It simply demands that the design and engineering department maintain documented procedures. This is certainly a basic quality system concept. The design and engineering department is to have a process in place for controlling product development (such as the ideal process already presented), and the procedures that are based on this process must be documented. These procedures could be called design standards, or standard procedures, quality procedures, etc. depending on the organization. As mentioned earlier, this system of documents needs to be reviewed at least at regular intervals.

Design plans

Paragraph 4.4.2 of ISO 9001 is entitled Design and Development Planning and covers the concepts of Design Plans and Adequate Resources. Design Plans refer to the activity of being proactive before a product development project is engaged. If the same type of product is always developed by an organization, then this may mean verifying that the proper preparations are made for the project. If the project is large-scale or unique in some way, then Design Plans could mean a very significant activity including planning and budgeting for new devices, new personnel, new procedures, etc. As with the process review suggested for these type of projects, the Design Plans could become another complete stage in the ideal process. However, in the author's view of the many types of products and organizations involved with ISO 9001, it is not certain that Design Plans must be considered part of the ideal process followed by every development project (particularly for smaller, technologically specialized departments or firms).

Adequate resources

In conjunction with the Design Plans, paragraph 4.4.2 covers the need to have adequate personnel and resources. This concept in paragraph 4.4.2 can be called Adequate Resources. If the Design Plans activity finds that there is not adequate personnel and resources (computer hardware and software, testing equipment, physical plant, etc.), then the Design Plans activity must dictate how that deficiency is to be addressed. As mentioned earlier, the Design Plans and Adequate Resources concept for the design and engineering department needs to be reviewed at least at regular intervals.

Organizational interfaces

Paragraph 4.4.3 of ISO 9001 is entitled Organizational and Technical Interfaces and covers the concept of interfaces. These interfaces occur between the various groups in the organization which are connected with the design process. These groups might be analytical, design, testing, manufacturing engineering, technical publications, etc. The interactions between these groups need to be defined and documented to satisfy ISO 9001. These interactions should be incorporated into the Documented Procedures concept presented earlier; therefore, the interactions should be defined in the design standards, operating procedures, quality manual, etc. The interfaces are probably most often documented through organizational charts and job descriptions. Again, as mentioned earlier, the Interfaces concept for the groups in the organization needs to be reviewed at least at regular intervals.

Revision control

Finally, paragraph 4.4.9 of ISO 9001 is entitled Design Changes and covers the concept of Revision Control. Design and engineering departments are likely to already be quite familiar with this concept. The Revision Control concept simply means properly managing the status or disposition of the design. As drawings and other documentation are created and evaluated in the course of the design process, the design is going to change. As drawings, for instance, are revised, they are given a revision level. Of course, some changes are made in the preliminary design stage and are simply changed. However, once a drawing is released or used in the production of a product, the changes need to be unambiguously defined, executed, reviewed, and rereleased.

Many organizations use a revision level concept to indicate the status of a drawing or document. However, different organizations use different systems to manage these levels and have different interconnections with various parts of the company (such as purchasing, production control, etc.). Typically, though, one finds that there is a Engineering Change Request or Design Change Request procedure for initiating a change to a product design. Subsequently, there is also an Engineering Release or Engineering Change Notice procedure for completing the design change. All these types of procedures are necessary for Revision Control.

Regardless of the specific method used for Revision Control, ISO 9001 requires that there be a formal procedure in place. This formal procedure must include an approval cycle, as well. Changes must be approved by authorized personnel, and such approval needs to be documented. Many organizations use an approval block in drawings, for instance, and these drawings contain a revision level and description block. Obviously, the approval for the changes are to have been completed before the changes are actually implemented. Failing to do so would indicate a lack of proper Revision Control.

Checklist

As shown in Figure 3.10, all these more general concepts for the design and engineering department can be made part of a checklist. This checklist can then be used to review the process for each product design project or at regular intervals. Again, this process may or may not be part of every product design project. If it is, then an entirely

---

Process Review Checklist

Procedures:

    Does the development project require new procedures?   Yes ☐ No ☐

    If so, which procedures need review? _____

Plans:

    Does the development project require special planning?   Yes ☐ No ☐

Resources:

    Are there adequate resources for this development project?   Yes ☐ No ☐

    If not, which resources (personnel, software, hardware, etc.) are inadequate?

    _____

Interfaces:

    Does the development project require special communications or changes in the organizational structure?   Yes ☐ No ☐

Revision Control:

    Does the development project require special revision or configuration management control?   Yes ☐ No ☐

Design and Development Process Review Form (sjs Rev 0; 8 March 1996)

Figure 3.10   Development Project Process Review Checklist.

new stage can be added to the overall ideal process as shown in Figure 3.11.

Conclusion

To summarize, a complete idealized product development process is shown in Figures 3.7 and 3.8. If warranted by the extent or nature of the product design being contemplated, process review activities can also be added as shown in Figure 3.11. Studying these materials along with paragraph 4.4 of ISO 9001 should prove helpful in bringing a design and engineering organization into compliance with the Design Control section of ISO 9001. As mentioned in the Introduction section of this chapter, there are many other parts of ISO 9001 to be concerned with, but the material presented in this section gets to the core of the issues for the design and engineering organization and its management.

Caveats

As a design and engineering department compares the ideal ISO 9001 to existing practices, it is certain that difficulties will be discovered and interpretations needed. Two areas addressed at this time are smaller organizations and information technologies. They are mentioned here since the remaining chapters that present a number of sample procedures and documents to implement the ideal ISO 9001 process tend to be geared toward the traditional paper-driven systems found in traditionally larger organizations.

Small organizations

One caveat or special problem is clear for small organizations without specific departments for activities such as design, analysis, testing, engineering documentation and standards, etc. In these organizations, the stages presented may not be distinct. In fact, the same people may perform different activities within the different stages for design projects.

In these cases where the stages are not organizationally distinct, the various checkpoints presented can be used as the delimiter. Indeed, each stage presented ends in a checkpoint or decision point. The Design Input Stage has the Input Review activity, the Design Stage has the Design Verification activity, the Design Output Stage has the Design Review activity, and the Design Validation Stage had the Design Validation check for customer requirements.

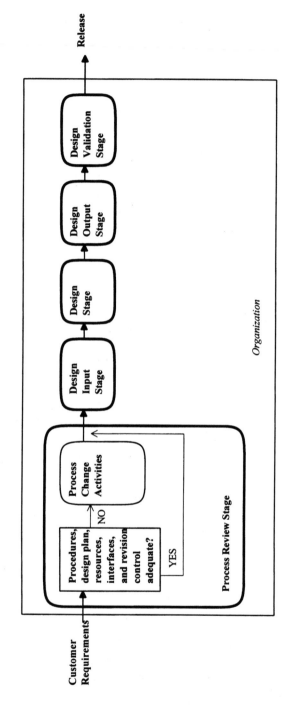

Figure 3.11 Ideal ISO 9001 with Process Review Stage.

If the checkpoint activities are performed, then no matter how blurred the organization, these points in the project should be distinctly identified and considered the boundary between the different stages. Of course, small organizations may move through these stages much more quickly and pass through the checkpoints more often during a project. However, as long as each occurrence of the checkpoint activities are documented and controlled, there should not be great difficulty in demonstrating that released designs have in fact met the design control demands of ISO 9001.

Another consideration for the small organization is having the same people perform design and then perform review design activities (such as the Design Review activity presented). There is no indication in the ISO 9001 standard paragraph 4.4 on Design Control preventing the same people from doing these activities.

Information age technology

Another special problem for the ideal ISO 9001 product design and development process is new paradigms flowing from Information Age technologies. For instance, meetings with minutes, two-dimensional drawings with approval blocks, physical prototypes, and even structured organizations may become things of the past.

Meetings (such as design reviews) can currently be held remotely via on-line conferences. The comments and ideas presented at such meetings may all be on-line and electronically recorded in computer files. In this case, the participants do not need to participate in the meeting simultaneously; instead, the discussion can flow over longer periods of time with opinions expressed and rebutted over days or weeks. An electronic data management system could then have the participants electronically approve the meeting computer files at the completion of the meeting to signify approval. In this case, there may be no paper trail to show to a quality system auditor. Of course, if a proper level of control over the computer system can be demonstrated, then the electronic data format and approvals may be acceptable to an auditor. In fact, the ISO 9001 standard does have a note that documents can be in the form of any type of media (including electronic media).[2]

Traditional drawings are also clearly called into question by the latest generation of computer technologies. It is now possible to economically define a part or assembly with three-dimensional computer models. As with meetings, the approval of these models can be done electronically. Again, this may not be a problem for ISO 9001, but care should be taken to demonstrate that the computer-based system is under an appropriate level of control.

Finally, some of the latest ideas about working in the Information Age deal with "virtual corporations." In this system, remote users or firms are linked electronically to perform a design task. These users or firms exchange all data electronically and create an electronic package encapsulating the design. After the package is created and passed on to the receiver or producer of the design, the organizational links may be broken. This scenario presents very serious problems for ISO 9001. How can documented procedures be controlled when the organization that performed the design does not even exist at the time of the audit?

Although the organization does not exist for the long term, the receiver of the electronic package could perform and document its own validation or inspection or testing once it is received from the virtual corporation. In this case, the design itself could be treated as incoming product such as raw material. However, in this case, the receiver of the package would not seek ISO 9001 registration. Instead, it would seek something like ISO 9002 which does not apply to design activities.[3] Thus the virtual corporation still does not seem to be compatible with ISO 9001. Although the virtual corporation concept is not the norm at the time of the writing of this book, it may become more popular in the future.

References

1. For the purposes of this book, ISO 9001 is equivalent to stating ANSI/ASQC Q9001-1994. The copyright date of this standard is 1994. The title of the American National Standard is *Quality Systems—Model for Quality Assurance in Design, Development, Production, Installation, and Servicing.*
2. ASQC, *ISO 9001 (ANSI/ASQC Q9001-1994) Quality Systems—Model for Quality Assurance in Design, Development, Production, Installation, and Servicing,* ASQC, Milwaukee, Wisconsin, 1994, paragraph 4.5.1 or note 15, p. 4, approved August 1, 1994.
3. ISO 9002 (ANSI/ASQC Q9002-1994) is entitled *Quality Systems—Model for Quality Assurance in Production, Installation, and Servicing,* Milwaukee, Wisconsin, 1994. This standard does not cover design activities.

4

The Design Input Stage

Introduction

This chapter begins the presentation of detailed information and examples on the workings of the stages presented in the previous chapter on the ideal ISO 9001 product design and development process. This information is intended to guide the reader in understanding the relationship between these ideal activities and a working design and engineering department of an organization. Hopefully, this is going to help the reader create or adapt his or her own department procedures to meet the requirements of ISO 9001.

In the Design Input Stage, the design parameters are established and documented. The design input documentation consists of technical specifications to be satisfied by the design. This information would then be passed into the Design Stage. As usual for ISO 9001-compatible processes, it is essential that the design input be documented in a specification document of some kind (written or electronic). In many organizations, this document is called a design specification.

The design specification must flow from the end-user or customer requirements. In cases where there is a contractual agreement pertaining to the product design, the design specification must also flow from that contract and its reviews. Obviously, it is assumed that the customer requirements have been understood, documented, and approved prior to the design and development process. One should note that while the customer requirements may be qualitative (e.g., design a stereo sound system that produces good sound), the design input would generally be quantitative and more useful for those involved with the design and development process (e.g., the design specification would dictate that an FM receiver is to be designed with a signal-to-noise ratio below a certain value). Or, one may consider that customer requirements are more macroscopic (such as compressing a flow of gas from one pressure and temperature to another pressure); while the design specification may be more microscopic (such as create a ruled surface centrifugal impeller with a maximum overall diameter and a maximum rotational speed and resulting stresses). The degree of detail and length of the design specification is going to vary widely between organizations and products.

Although some small organizations may not consider the Design Input Stage to be a significant expenditure of time and resources, many firms are going to spend a significant amount of time and effort in the Design Input Stage interpreting and understanding the customer's needs. For instance, large-scale or multiyear customized design projects with a specific contractual agreement may require detailed analytical calculations (such as performance predictions and environmental studies) before any design work is attempted. These calculations may be performed over long periods of time within the Design Input Stage. Therefore, this book assumes that the Design Input Stage is a significant part of the design and development process.

In some cases, it may be necessary to revisit customer requirements, review contracts, or revise design specification documents due to technical issues raised by technical experts in the Design Input Stage. Thus, the Design Input Stage is generally going to be tied into activities outside the design and engineering department. These activities are coordinated with marketing, customer relations, contract administration, or legal departments. This aspect of the Design Input Stage must be accommodated by the procedures used in the Design Input Stage and must be allowed for by the design specification documentation.

This Design Input Stage may be considered as overhead or a loss in relation to actually doing design work. However, the Design Input Stage is essential to attaining quality in design (which is what ISO 9001 is supposed to help bring about). The earlier and more often customer requirements are reviewed and discussed within the organization, the more likely the final product is going meet those requirements on time and within budget (assuming the management of the organization is capable of efficiently achieving consensus agreements). Few projects are able to withstand being resumed after significant design work is complete because the design input documentation was wrong or ambiguous or misunderstood.

In spite of the effort spent in achieving a stable design input document, revisions may be required and must be controlled. For example, some design or engineering departments may be pushing the state of the art with a given design. After performing simulations or performance predictions within the Design Input Stage, it may be found that a requested specification is not attainable. Therefore, there must be an ability to revise and/or refine the design input documentation. The most likely vehicle for the control of this process is the creation and control of a design specification document and the application of revision control to that document.

Although the design input document or documents may take various forms, some items must be considered for inclusion on this documentation based on ISO 9001. In particular, ISO 9001 states that the design input documentation is to identify applicable statutory and regulatory requirements. Depending on the industry or product, there could be a checklist of possible industry standards and/or government regulations to be considered. Also, ISO 9001 states that the design input documentation must be reviewed for adequacy. Thus, an approval signature is added to the document by a person given the responsibility for reviewing the documentation.

Another class of information that needs to be part of the design input documentation is acceptance criteria. These are specific targets or requirements that are to be met by the design. This information needs to be in the design input documentation so that it is readily available in the Design Output Stage (where these targets are reviewed for the design).

Design Specification

Figures 4.1 and 4.2 show some examples of design input documentation (or design specifications). The first example is meant to apply to a large project, while the second example is meant to be for a small project or organization. In both cases, there is no sample technical design specification information shown. Instead, the components and format of the documents are presented as typical document organization.

Figure 4.1 (the example for a large project) actually shows a master document page that acts as a cover page for a larger document. Then there are entire subsections devoted to components of the design specification. The master document is meant to provide review and approval for the entire design specification package.

The heading information serves to clearly identify the project that is being specified. The example shows a project number type of designation and a project name. Preferably the project number or other designation is then listed in a master list (or Product Development Master Log as shown at the top of Figure 4.1). This master list provides a clear level of control over the organization's past, current, and pending development projects. This procedure should also allow the

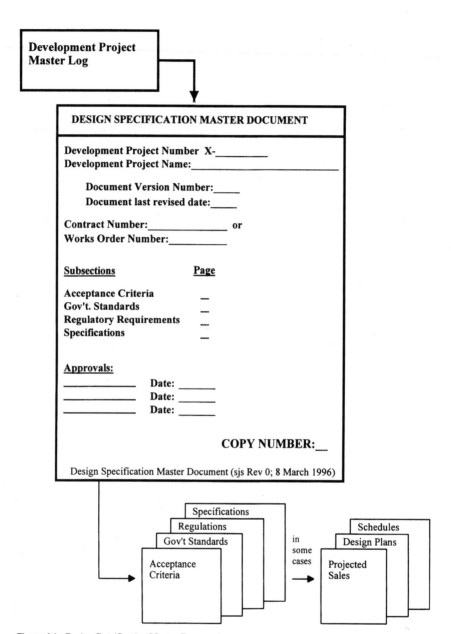

Figure 4.1   Design Specification Master Document.

organization to consider the design specification document as controlled since it is recorded in a master list. Furthermore, the document shows a version number near the top and a copy number at the bottom.

If the design specification is revised, the version number provides an indication of the disposition of the document. This version number would then also be listed in the master list (to again allow it to be considered controlled). Finally, assuming the design specification document is distributed to various people, departments, and groups, the copy number can be used to uniquely identify each copy. Again, the copy numbers and the recipients can be listed in the Product Development Master Log master list to make the design specification a controlled document.

Clearly, in this type of project it is going to be essential that everyone in the project have the current and correct design specification document. Obviously, customer satisfaction would be at risk if different parts of an organization (marketing and engineering, for instance) assume different design specifications to be true. Controlling the design specification document by listing it in a master list with copy numbers and version numbers should serve to prevent this problem. When designers, engineers, or others need access to the design specification information, one of these controlled documents should be consulted directly. This is the only way to ensure that the information is current and correct.

If some or all of the design specification document is copied, then it must be identified as a reference copy. In many cases, this is done by a stamp on the document indicating "for reference only." However, reference copies should be avoided by all means. Of course, if the information is created and maintained in electronic format, the information could be placed on-line. In this case, all those involved could consult the live document and would thus be more likely to have the correct information.

As Figure 4.1 shows, the master document contains some typically pertinent information such as a Contract Number or Works Order Number. Subsequently the document lists the ISO 9001-required items such as acceptance criteria (items which must be met by the design by contractual agreement, market critical design factors, performance guarantees, etc.), government standards to be applied to the design [ISO, ANSI, ASTM: American Society for Testing and Materials, ASME, American Nuclear Society (ANS), Institute of Electrical and Electronics Engineers (IEEE), etc.], local or national regulatory requirements [underwriting tested marks, local inspections or codes, ASME Boiler Code stamps, European Community (EC) directives, etc.], and finally technical specifications (design parameters, performance envelopes, etc.).

Even if requirements such as regulatory requirements are often assumed for an organization's products, they should still be listed for every project in the design specification. One cannot tell when a new employee or other person reads the design specification with no prior knowledge of the organization's procedures. Since these regulatory requirements are so vital, they should simply be repeated on each design specification document issued. Electronic documents can be used to great effect by easily keeping this information on a template without requiring the user to enter the basic regulatory requirements repeatedly for each development project.

Since the ISO 9001 Design Input section requires that Design Input Stage documentation be reviewed as well as controlled, an approval section is added to the sample design specification in Figure 4.1. These signatures should be placed under a statement to the effect that the undersigned have reviewed and approved the entire design specification document. The approval should signify that the design specification is complete and unambiguous, with all the responsibilities for the design activities assigned and all the adequate resources available. Furthermore, if there is a contractual agreement for this product design, then the approval signifies that the design specification is in accordance with that contract and the results of any reviews of that contract.

Everything presented in the master-type of design specification document has been geared toward success with audits. All the information needed to demonstrate conformance with the ISO 9001 section on design input is clearly shown. Assuming proper control over this document can be demonstrated, an audit for design input should be successful. The last information to note in Figure 4.1 is the form Identification at the bottom of the sample document, "Design Specification Master Document (sjs Rev 0; 8 March 1996)." Just as control over the design input needs to be demonstrated, control over this form also needs to be considered. If the form is changed by management of the organization, then all those involved with design input procedures would need to be aware of the changes. The form identifier (form name, electronic document name, etc.) and the date of its release should provide this control.

Figure 4.2 shows a design specification document for a smaller design project. This document may be more suitable for smaller organizations or for organizations that design the same product with small variations. This form may also be suitable for larger organizations that require a streamlined procedure for more immediate results.

```
┌─────────────────────────────────────────────────────────────┐
│  DESIGN SPECIFICATION                                        │
├─────────────────────────────────────────────────────────────┤
│                                                              │
│  New Product Code Number:_____        Date:_____       │
│                                                              │
│  Design Specifications:                                      │
│        1:_____           │
│        2:_____           │
│        3:_____           │
│                                                              │
│                                                              │
│  Regulations:                                                │
│        1:_____           │
│        2:_____           │
│        3:_____           │
│                                                              │
│                                                              │
│  Safety Requirements:                                        │
│        1:_____           │
│        2:_____           │
│        3:_____           │
│                                                              │
│                                                              │
│  Approvals:                                                  │
│  _____    Date: _____                             │
│  _____    Date: _____                             │
│  _____    Date: _____                             │
│                                                              │
│                         COPY NUMBER:__                       │
│                                                              │
│                                                              │
│  Brief Design Specification (sjs Rev 0; 8 March 1996)        │
└─────────────────────────────────────────────────────────────┘
```

Figure 4.2   Brief Design Specification.

The sample form in Figure 4.2 begins with a new product code as a sample project identifier. Each time one of these forms is completed, a number would need to be assigned. If all these sample forms were completed by a single person in a marketing department (for a smaller organization perhaps), then the projects could be simply numbered sequentially. A name of a proposed new product might be sufficient for identifying a new development project, but a system would need to be developed to prevent reuse of the same name, and prevent confusion. Otherwise members of the organization would not know the disposition of a particular project (if the same project name was accidentally used). As with the more complex document in Figure 4.1, control over the activity in the Design Input stage must be maintained. Of course, if there is a contractual agreement that applies to the development project, then the contract needs to be referenced and any contract review must be reflected in the design specification document.

The next section of the form (after the heading Number and Date) in Figure 4.2 is for the Design Specifications. In the case of an organization that develops very similar products, this section could even have check boxes for different parameters of the same basic product and blank lines or boxes for the desired values for the different parameters.

The next three sections correspond with the specific requirements of the ISO 9001 standard for the Design Input Stage—acceptance criteria, regulatory requirements, and safety requirements. As stated earlier, this specific information is going to be needed in the Design Output Stage. For the simple projects implied by Figure 4.2, the regulatory requirements and safety requirements may be the same for all these projects. In this case, the same requirements could be made part of the form. However, a check box next to each should be provided so that the person completing the form could then check off (or tick) each regulatory requirement. This method is a better demonstration that the needs of the particular design have been considered.

At the bottom of the form in Figure 4.2 is a section for approval signatures. With all the information for the design specification on the same form, the signatures and dates should then show that the information on the form is complete and unambiguous. If there is any question as to the intent of the approval signatures, the procedure which covers the use of the design specification form (in the Quality Manual or the departmental procedures) could state that those approval signatures signify that the person signing the form has reviewed the information and considers it complete and unambiguous.

Figures 4.1 and 4.2 present some examples of the design specification documentation. This documentation becomes the marching orders of the activities of the Design Stage, which follows the Design Input Stage. As a final check on the correctness, adequacy, and consistency of the design specification, an Input Review meeting activity could be made part of the standard design and development process. At this meeting, the contents of the design specification documentation could be compared to any available contractual agreements. If for no other reason, this meeting can signify the end of the Design Input Stage and the beginning of the Design Stage (particularly for smaller organizations). If a meeting procedure is implemented, then minutes could be kept as an appendix of the design specification. If there are revisions to the design specification in light of the Input Review meeting, then the activity of the Design Input Stage should continue until the issues are resolved.

Another possible aspect of a design specification or new product proposal is projected sales. In fact, this aspect highlights a significant difference between "engineered products" and "marketed products." In the case of the engineered product, most likely an order is received for a specific product to be developed or adapted for a specific customer. That customer order is already going to have some pricing included so that the product can be delivered to the customer and a certain amount of money can be received.

By contrast, in the case of the marketed product, this advanced financing does not really exist. Although some customers may place orders for a marketed product before development, most likely the organization is developing the new product based on a predicted number of units being sold within a certain amount of time. Obviously, the pricing of the product is going to be somewhat based on the design parameters chosen in the Design Input Stage; therefore, the projected or predicted sales may be a logical addition to the design specification. If the marketed product's projected sales is to be part of the design specification documentation, then the appropriate members of a sales and marketing department needs to be included in the Design Input Stage. They should participate in the approval of the documentation as well, and they would need to be present at any Input Review meeting.

Having considered whether or not to include projected sales in the design specification, the reader should note that the ISO 9001 standard often seems to apply best to the engineered product (at least if one assumes that the engineering product is obtained by a contractual agreement). Paragraph 4.3 of ISO 9001 (the paragraph that precedes Design Control) is devoted to Contract Review. Although it may then seem correct to assume that ISO 9001 only applies to engineered products with a written contract, this does not seem to be the case.

Finally, then, although the need for projected sales information in the design specification is not considered by ISO 9001, there should be no problem with including it and using the design specification document as a means of approving the risk of new product development in light of perceived market conditions. Whether or not the product succeeds financially in a market is not the primary concern of ISO 9001. The primary concern is to see that the customer's requirements are met on the first attempt. Although meeting a market's needs efficiently should have a positive financial effect, there is no guarantee of this. For example, poor sales and marketing strategies cannot be prevented by ISO 9001.

Other aspects to be considered at the Design Input Stage are scheduling and resource allocation. Although an acceptable set of design parameters may be established for a new product, the development project may not be approved due to a lack of funding, personnel, laboratory space, etc. This may be particularly a problem in organizations where multiple development projects are occurring simultaneously. These projects may be competing for existing resources and the priorities of these projects may be fluctuating.

Design Input Flowchart

To summarize the Design Input Stage, consider the flowchart presented in Figure 4.3. This flowchart attempts to show what might happen within a typical organization during a Design Input Stage. Prior to reaching this stage, the example in Figure 4.3 shows some preliminary activities such as market research. These are marketing-

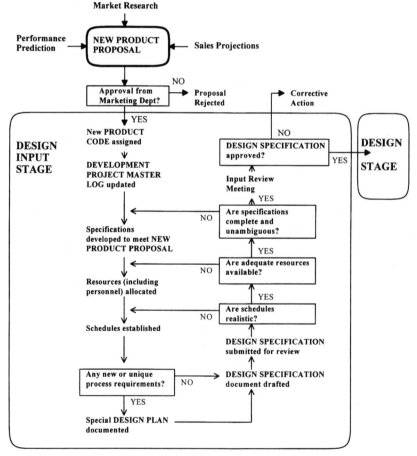

Figure 4.3   Design Input Stage sample process.

related activities assumed to be outside the scope of the Design Input Stage.

In this example, a marketed product is assumed, and the first step shown is market research leading to the creation of a new product proposal. This proposal could be a marketing-controlled document similar to the design specification controlled by the engineering department. This proposal would contain the customer requirements as understood by a marketing department. Of course, for the engineered product case, this part of the process would be different.

Figure 4.3 shows that there are some significant inputs to the product proposal. One such input is called performance prediction. This signifies the various calculations that are likely to be done during the proposal stage. Since this overall process is for a new product development project, and since the new product may contain variations or innovations never designed before by the organization, there is likely to be a need to predict the behavior of the product (without actually creating a prototype). In this marketed product example, it is assumed that some performance prediction is required.

The second input to the product proposal is sales projections. This concept has been presented earlier as being included in the Design Specification document. However, the author assumes that this type of information is generally initially developed in the marketing department of an organization. Therefore, this information is best considered part of the product proposal document (at least as it is first projected). The projected sales data can then be easily included in the design specification later, and the projected sales information can then be controlled from that document once the Design Input Stage is reached. Making the initial projected sales part of the product proposal also gives the proper timing for the next step in the example.

The next step in the example in Figure 4.3 is approval. At this step, the marketing department of the organization is assumed to make the decision on whether to proceed with the new product development. Certainly at this point, the person or persons making the decision need(s) to see the projected sales information as well as the performance prediction. The decision needs to consider the competitive advantage of the new design (based on the performance prediction), and then determine if the organization's investment in this project is going to be paid back by the projected sales (over some amount of time). Of course, this is a subjective decision. The cost to build the new product cannot be completely determined until the design is completed, but the Design Input Stage has not even been started at this stage. Nonetheless, the project's effort should be approved at this stage based on whatever information is available, since the Design

Input Stage is going to be initiated next. Since it is assumed that the Design Input Stage is going to absorb significant resources of the organization, there should be an approval to proceed, and the purpose of the marketing approval step is to provide that approval.

One should note that although the marketing department of the organization is assumed to make the decision to proceed in this example, this decision-making process could easily include the design and engineering department or the financial department. The amount of approvals required should reflect the investment to be made by the organization (in terms of financial impact) or in terms of the number of hours of effort (within the design and engineering department).

As shown in Figure 4.3, the approval step for the new product proposal could result in a rejection of the proposal. That is, the management of the organization (marketing, engineering, finance, etc.) may decide that the proposed new product does not have an attractive payback, or it entails too much risk, etc.. At this point, the new product proposal could be reworked or resubmitted. Alternatively, the new product proposal could be dropped altogether.

Of course, the new product proposal could also be accepted. As shown in Figure 4.3, the process has now entered the Design Input Stage. The design and engineering department would now devote some effort to the project. The first step shown is the assignment of the new product designation or code. The new product proposal may have already defined the name or designation of the product, but the design and engineering department should record the designation for its own records. Also, there may be new product designations which do not proceed past the marketing approval. The new product name (to be seen by the market) may also not be established yet at this point in the process. Therefore, a code number or name should be assigned by the design and engineering department for its own use.

The design and engineering department needs to be certain which projects are in the Design Input Stage since engineers from the department may assist in or perform the performance prediction calculations mentioned earlier (this activity is outside the Design Input Stage). For this reason, the next step in the sample process shown is to update a Development Project Master Log. This log document is a master list of all the projects which have been undertaken by the design and engineering department (and that have reached at least the Design Input stage). Figure 4.4 shows a possible format for the Development Project Master Log (which is also seen in Figure 4.1).

The next step in the process shown in Figure 4.3 is actually developing the design parameters or specifications for the new product. Obviously, this step takes account of the new product proposal information. The design and engineering department should now know

DEVELOPMENT PROJECT MASTER LOG				
Entry Date	New Product Code	Reference Proposal	Estimated Product Release Date	Actual Product Release Date

Page:_____

Development Project Master Log (sjs Rev 0; 8 March 1996)

Figure 4.4 Development Project Master Log.

what the marketing department expects in terms of performance. The technical data must now be specified to reach that level of performance. This data then become the basic technical input to the designers and engineers in the Design Stage. Details of this part of the process can hardly be presented since this is a fictitious case. Indeed, the reader is expected to know what technical data are relevant to his or her organization.

The next steps in the process are performed by the management members of the design and engineering department. These steps consist of determining resource and personnel allocation and establishing schedules. At this point, management would be looking at the current workloads, the scope of the new product development project, and the priority of the new project versus existing projects. It is assumed that the resources and schedules being developed are for the Design Stage, not the Design Input Stage (which is assumed just to be a burden or overhead). Engineering management would typically look at the people, software, hardware, laboratory facilities, prototype build resources, testing resources, etc. that are going to be needed during the Design Stage where the design is going to be realized.

Design Plans

The resource allocation and scheduling tasks currently being discussed are related to the concept of Design Plans presented in Chapter 3 on the ideal ISO 9001 process. As mentioned in that chapter, the nature of the product and/or organization may warrant an entire separate stage devoted to Design Plans. This would probably depend on the size of the new product development projects and the amount of variation or innovation in the new products. In the current example, it is assumed that most of the new products being developed are basically variations of the same product (the case of developing new products in a product line). In this case, the needs of the development project are basically known (although one should assume that there may also be special cases or exceptions).

To create a contingency, then, for unique projects for which the basic procedures of the organization are not sufficient, the next step in the process in Figure 4.3 is a check or decision about the project's needs. This checkpoint asks if there are any unique requirements for this particular project. If engineering management feels that unique requirements exist, a Design Plans activity can be initiated. In this activity, the unique requirements are identified. Subsequently, plans must be developed to meet each new requirement, and the plans must consider the possible problems likely to be encountered. The Design Plans should contain contingencies for these unique requirements since this is likely to be new territory for the organization to consider. Finally, it is assumed that the documented results of the Design Plans activity are included in the design specification documentation, which is being developed in this Design Input Stage.

The next step in the sample process (after checking for the need for Design Plans) shown in Figure 4.3 is the creation of the design specification document. Some possible formats for this documentation have already been presented earlier in this chapter (see Figures 4.1 and 4.2).

After the design specification documentation is completed, it can be submitted for approval in the next step. A number of questions should be covered in this approval process. As shown in Figure 4.3, the following are sample questions:

1. Are the schedules realistic?

2. Are adequate resources available?

3. Are the design specifications complete and are they unambiguous?

A negative response to one of these questions is shown to lead to a resubmittal to one of the activities already covered (establishing design stage schedules, allocating personnel and resources, and developing design specifications).

In the sample process shown, after all the questions posed above have been adequately addressed, there would be an Input Review Meeting. At this meeting all the management personnel responsible for approving the design specification would meet and review the design specification and decide if the design specification is acceptable. If problems are determined to exist at this point, Figure 4.3 shows that Corrective Action should be initiated. This action could be changing the process or the personnel involved. Or it could also be to return the project to the marketing department due to problems with the new product proposal.

Assuming that there actually are no problems at the Input Review meeting, the management personnel would sign the design specification document, and the meeting's minutes could be included as an addendum to the design specification. This meeting would then serve as a clear ending point of the Design Input Stage. At this point, the design specification can be moved into the Design Stage according to the schedule included in the design specification.

Conclusions

This chapter has presented some details concerning a Design Input Stage that is perceived to be part of an ideal process for ISO 9001 demands. The Design Input Stage activities are centered around the creation of a design input document such as a design specification. This document needs to translate the customer's requirements into specific design parameters. Although the Design Input Stage is likely to be considered overhead, it needs to be given proper attention since changes to the design specification once the Design Stage is entered will have a large negative effect due to rework. It would be most efficient to invest the effort in the Design Input Stage toward improving the chances of doing the Design Stage work right the first time.

5

The Design Stage

Introduction

The Design Stage is where the product is actually designed or invented. Although the audience of this book is probably familiar with what must happen to design a product, some general comments about designing are presented here to guide those less familiar with this activity. A momentary reflection on this activity might also benefit designers and engineers who need to refocus on the activity from an outsider perspective. This is the proper perspective for understanding the ideal ISO 9001 design and development process being discussed since the customer is assumed to be an outsider or external to the design and engineering organization. Subsequently, some unique difficulties of ISO 9001 versus design-related and innovation-related activities are discussed.

There are many different opinions on what designing means. To some it designates the creation of completely new artifacts for practical application (i.e., invention). To others, designing means the synthesis of existing knowledge and new processes to adapt existing artifacts (i.e., a more evolutionary process). To yet others, designing would be a combination of these ideas.

For the purposes of this book, designing is simply considered the act of determining the configuration of a product. Furthermore, it is assumed that the Design Stage is the only stage where configuration is determined (i.e. what a product looks like, how it functions, what raw or processed material it is made from, and how it is assembled). Although activities such as manufacturing and field operations may precipitate changes in a design, for this ideal ISO 9001 process, these changes are assumed to be channeled through the design or engineering department for implementation and approval. This approach should also be able to accommodate concurrent engineering needs by simply including the various disciplines into the design and engineering environment and having the design and engineering department direct the Design Stage activities. Thus, it is only the Design Stage that determines the configuration of the product. Note that in view of this interpretation, all other stages in the design and development process, as well as all other processes in an entire organization could be considered enabling functions in support of the activity of design.

The design-centered view of the entire organization seems to be reinforced by ISO 9001 which emphasizes the importance of quality in design.[1] ISO 9001 emphasizes that if the Design Stage fails, all other processes are going to fail, and customer needs simply cannot be fulfilled. This conclusion should be accepted by designers and engineers. They should realize how vital they are to the process and the organization. Having accepted the importance of their contributions, hopefully designers and engineers can accept the challenge of the procedures and policies that are going to be created in their departments in order to meet ISO 9001 requirements.

ISO 9001 Challenges

There are a number of difficult issues to initially address in the Design Stage. One such difficulty is that a standard such as ISO 9001 obviously cannot tell an organization how to design its product. However, there is a requirement to maintain control over the design procedures. This may not be difficult in some cases, but in other cases too much control may stifle innovation. Therefore, the difficult issue of control versus innovation arises. Each organization is going to have to deal with this issue, and will hopefully come to a consensus-based agreement on the level of control and documentation most likely to meet ISO 9001 with a minimum of interference with innovation.

Some other issues that may need to be confronted when viewing ISO 9001 in a design activity are the use of trial-and-error methods, individual responsibility versus design teams, secrecy, and the proper use of reference and other uncontrolled documents. These issues have some relevance to the ISO 9001 standard, and should be given consideration when a design and engineering organization considers ISO 9001 registration.

Trial and Error

In a practical sense, the design activity almost always involves a trial-and-error activity (although it is hopefully guided by past experiences). There really is no case where everything is completely new in a newly invented product manufactured using never before imagined material processing procedures. Instead, the designer proposes a real solution to the problems posed by the design input documentation. This proposed solution is likely to be based on the designer's previous experiences. The proposed solution is then tested in a variety of ways.

The proposed solution may be tested by making theoretical calculations. The proposed solution may be tested by making empirical (nontheoretical) calculations or comparisons. The proposed solution may be reviewed by experts or personnel with field experience. The proposed solution may be tested in a laboratory or other environment. In each of these cases, the designer is searching for a reason the design is not feasible, or the designer finds confirmation that the design is feasible. At the same time, other possibilities may present themselves. If the proposed solution withstands the reviews and tests, then it may be taken to a prototype stage. If the proposed solution fails some of the investigations, then the proposed design is modified or discarded altogether in favor of another solution. Thus, it appears that the basic design activity is based on a trial-and-error process.

Of course, trial-and-error processes may be seen by quality professionals as being at odds with Total Quality Management (TQM) or "zero defect" philosophies. However, designers and engineers must communicate the idea that one cannot avoid errors in probing the limits of the physical universe. These limits present unavoidable uncertainties. Like it or not, the progress of a design is going to be dictated by a designer's skill at formulating images or ideas of what can actually be created in the physical universe. No amount of TQM philosophy is going to prevent all setbacks in the designer's pursuit of solutions.

In any case, ISO 9001 does not appear to demand a "zero defect" process for the Design Stage. ISO 9001 assumes that the design and development department knows how to perform the design activity, and the proof of that competence lies in the performance of the product designed. Just the same, some organizations may have to deal with a clash between quality control personnel searching for a zero defect process and designers attempting to work with physical and performance uncertainties.

Of course, each of the trials and difficulties that are part of design activities presents an opportunity to learn and expand the knowledge base of the organization. If these investigations are properly documented, then they can be referenced for future efforts. Also, for ISO 9001, this documentation can be used to demonstrate an appropriate level of control over the Design Stage. This chapter is going to present a number of sample documents for containing this knowledge such as Design Files, Engineering Standards, Verification Studies, and Engineering Test Reports. In many cases, these types of documents are not given the attention they deserve during demanding commercial product development cycles. Hopefully, an ISO 9001 registration effort can be used to heighten the awareness of their value as a long-term knowledge base.

Design File

One possible format for documenting design activities through the fluid changes of the Design Stage is a Design File. A sample is shown in Figure 5.1. Each designer or engineer involved with a particular new product development project could keep their own file. The Design Files can be uniquely identified by a development project number or a new product code (as shown in Figure 5.1). Sections of this file would be devoted to items such as calculations, sketches, laboratory results, literature search, and reference materials. Considering the importance of the calculations and sketches of the

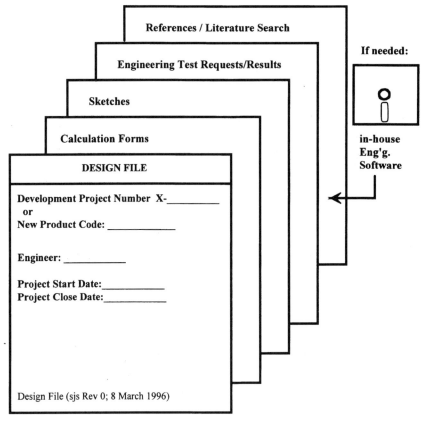

Figure 5.1 Sample Design File.

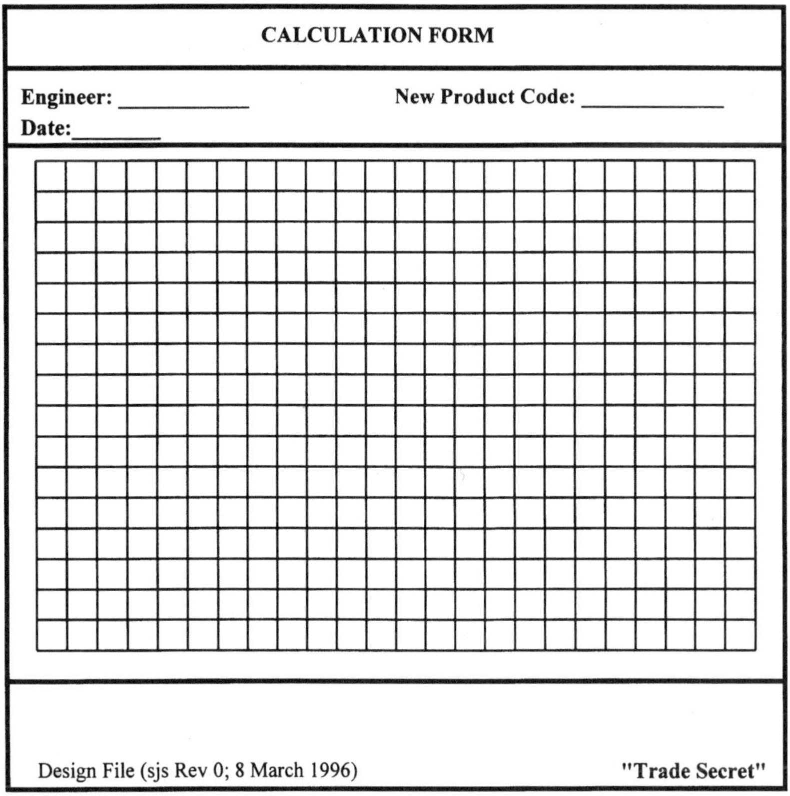

```
┌─────────────────────────────────────────────────────────────────┐
│                      CALCULATION FORM                             │
├─────────────────────────────────────────────────────────────────┤
│  Engineer: _____        New Product Code: _____       │
│  Date:_____                                                     │
├─────────────────────────────────────────────────────────────────┤
│                                                                   │
│                                                                   │
│                  [grid]                                           │
│                                                                   │
│                                                                   │
├─────────────────────────────────────────────────────────────────┤
│  Design File (sjs Rev 0; 8 March 1996)            "Trade Secret"  │
└─────────────────────────────────────────────────────────────────┘
```

Figure 5.2   Sample Calculation Form.

Design File, controlled forms should be used for these parts of the Design File. Figure 5.2 shows a typical controlled calculation form, and some formats for sketches are presented in Figures 5.18 and 5.19. As the design process progresses, parts of this file may be moved or copied into controlled documentation as part of the design output or as part of the design verification process. The intent of these forms is to show a level of control over the design activities, while allowing the iterative nature of the calculations and tests to move along efficiently.

Note that the whole design activity may take place within a single individual for some organizations. In other cases, it may involve entire departments, teams of people, or even entire organizations. One person may propose the design, while others may make various calculations to test the design. Some designs may not need analytical calculations and may be based entirely on past experience. In this case, design decisions may be made by designers. Other designs may require sophisticated new algorithms on computers, and in this case, the design decisions may be made by engineers with advanced degrees. The difficulty presented here is that ISO 9001 still requires the process to be documented and controlled to some degree. However, one still needs to allow the solutions to design problems to emerge from different parts of the organization without restricting the innovation process. In the best outcome, procedures and policies can be found to satisfy the needs of ISO 9001 while not restricting innovation. If not, the organization is going to have to judge the merits of ISO 9001 versus the loss of creativity.

Engineering Standards

The balance between control and innovation may be assisted by the development and use of design or engineering standards. These design standards are internally developed standards for guiding the design process. They are usually a numbered and controlled set of documents that contain various calculation methods, material selection guidelines, margin of safety policies, etc.

Engineering standards are useful for solidifying knowledge gained by the organization, and are generally proprietary. They are also quite useful for disseminating engineering methods for use by nonengineers. That is, as new calculation methods or algorithms are developed by analysts, new design procedures can be documented for use by designers. Assuming the engineering standards are properly controlled and capable designers properly trained, the designer could produce results with adequate accuracy. Of course, the limitations of the methods would need to be clearly defined. In addition, some analyses are not appropriate for this method and are going to have to be performed by specialists. The use of these engineering standards does have some bearing on an ISO 9001 registration effort.

Engineering standards are certainly valuable in the context of ISO 9001. They can show an appropriate level of control over the activities of the Design Stage. However, the organization must take care to show that they are controlled documents. The organization must keep a catalog of all the engineering standards (most likely through a naming or numbering system). The disposition of each of the standards must be known at all times (in process, active, in revision, etc.).

The copies of the set of standards must also be controlled. There must be a sequence number on the standards, and there must be a master list of all the copies and where they are located. If there are revisions to engineering standards, then they must be automatically forwarded to each copy holder for update. The electronic format can certainly assist with this type of control. If the documentation is in a document retrieval system on a computer network, for instance, then the standards can be automatically updated for all users of such a system.

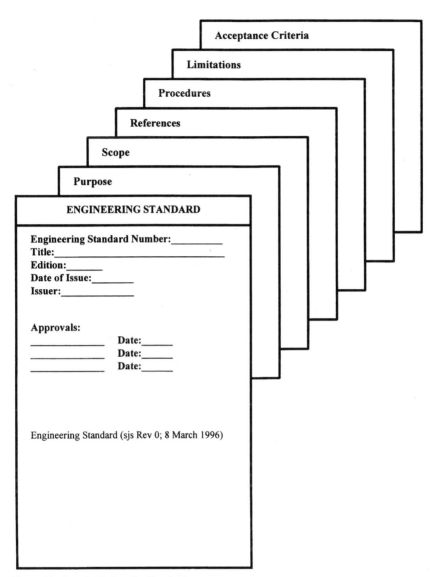

Figure 5.3 Sample Engineering Standard.

Figure 5.3 shows a typical format for an Engineering Standard. The important control information is at the top of the first page of the form. This includes the title of the standard, the revision level of the standard, the date of issue of the standard, and the name of the person who issued the latest revision. All this information is essential for demonstrating that the information is controlled. If an ISO 9001 audit finds that the latest engineering standards information is not available to the proper people (such as designers) in a timely manner, the organization is likely to fail the Design Control section of ISO 9001.

In addition to having the proper format (such as a form in Figure 5.3) and having the timely dissemination of the engineering standards, they must be approved by authorized personnel. Although this ISO 9001 requirement is under the subparagraph (4.4.9) on changes to designs, it can be interpreted to apply to design methods such as those found in engineering standards. Thus, the organization should implement an approval process for the issue and revision of engineering standards.

There are at least two possible methods of demonstrating the approval. In one method, each revision of the engineering standards could be signed by the authorized personnel. In another method, a signature form system could be used. In this case, an engineering standard would not be issued by the controlling department until a standardized form is signed. These forms with signatures would then be cataloged as quality records that can be retrieved during an audit.

Referring to Figure 5.3, one can see that the standard can follow a format that includes various sections. The first heading or section should be the Purpose. As implied by the name, this standard should state the purpose of the standard. This must be unambiguous versus other standards in the system. If there even appears to be more than one standard applying to the same calculation in the same situation (based on the title and purpose of the engineering standard), then there has arguably been a collapse of design control. It is essential to ensure that the designers are using the correct calculations. Furthermore, the author of the standard should realize that the person making the design judgments may decide to use one Engineering Standard versus another standard based solely on the title and the Purpose section or paragraph. Certainly, the use of catch-all or over-generalized titles should be avoided.

The next section shown on the sample Engineering Standard in Figure 5.3 is Scope. This is a widely used term to indicate the area or range of application of the standard. The Scope could indicate which groups or departments are to follow the standard. The Scope could indicate what kind of processes are applicable to the standard. The Scope could indicate which products or product lines the standard applies to. All this information should assist the reader in understanding whether the standard applies to a given design situation. If the standard applies to all areas or all products, then it should state this as well.

Following the Scope section, Figure 5.3 has a section for References. This is also a widely used term to indicate what other standards are relevant to a design situation. This could include names of industry standards (such as ASME, IEEE, ASTM, etc.). This section could include names of paragraphs of ISO 9001 as well (such as "ISO 9001 Paragraph 4.4 DESIGN CONTROL").

The remaining three sections in the sample Engineering Standard are assumed to apply to the design or calculation information contained within the standard. First is the Procedure section. Obviously, this section presents the calculation or logic to be used by the designer. This section is likely to include formulas for calculations, tables of data, charts for looking up data, flowcharts, "decision" rules, etc. One aspect of the Procedure section to consider are the unit systems (SI, customary, CGS, etc.) and whether they apply to local or global divisions of the organization.

Another situation to consider is the use of computer programs or spreadsheets for the calculation procedures. In this case, although the design method may be approved by the authorized personnel, the computer program introduces another possible area out of control. If Engineering Standards calculations are going to be performed by computer programs, these programs must be validated to ensure that the current version of the computer program matches the current version of the standard. One can refer to other works on the issue of engineering software management and quality control.[2-4] To summarize those works, if the computer programs cannot be controlled as well as the standards documentation, then the programs should not be used.

The next section of the engineering standard that applies to the calculation method is called Limitations. Particularly with the advent of computer programs, the ability to execute easily the calculations presented in an Engineering Standard is always being enhanced. With the results of the calculation readily available, then, the standard can then emphasize the interpretation of the results. The Limitations section should cover this interpretation. This section should clearly state when the results are to be considered invalid; it could also state when experts should be consulted to properly interpret the results.

The final section of the sample Engineering Standard format shown in Figure 5.3 is the Acceptance Criteria. Although acceptance criteria are discussed earlier with respect to the design specification and its eventual check against the Design Output Stage, acceptance criteria should also be spelled out within engineering standards when appropriate. These acceptance criteria apply to the results obtained from the methods presented in the standard. For instance, once a value is calculated by a formula in a standard, and it is considered a valid result by the Limitations section, then the question arises whether the value is within the margins for use in the product being designed or within the margins dictated by the design and engineering department (such as acceptable margins of safety). This section is perhaps most crucial in demonstrating control over the design process. The

Acceptance Criteria section is intended to show that safe, economical, practical products are going to be produced by the organization because of its experience with these calculations and its ability to keep them within the Acceptance Criteria.

Design Verification

The obvious question to be posed with respect to these Engineering Standards is how does one know that the methods are correct? In the author's opinion, this is a most important question associated with ISO 9001 for the design and engineering department. Just because a manager signs an approval form for an engineering standard does not necessarily mean that the methods and formulas it employs are any more likely to be valid models of the physical universe. Although some organizations may have qualified experts to review and approve Engineering Standards, in many cases a manager is going to simply sign the standard. Furthermore, internal and third-party audits for ISO 9001 are not going to question the technical competence of a manager beyond a level of formal education.

Although a manager's approval may be sufficient for demonstrating an appropriate level of Design Control for ISO 9001, the author feels that the pursuit of an effort such as ISO 9001 within a design and engineering organization should be used to foster a higher level of critical review of design methods. In particular, the Design Verification section of ISO 9001 can be used to justify a serious program to test and verify design methods of all kinds. That is, it is the author's interpretation that Design Verification can apply to Engineering Standards as well as specifically designed products or design projects. All too often, if one asks a designer or engineer for a direct comparison of an in-house algorithm or design method versus measured values (such as strain gauge results), one often gets a blank stare or the standard response that they don't have time for such things. In this cases, ISO 9001 can optimally be used as a justification to remedy this situation.

If Engineering Standards or design methods are encapsulated in computer programs that are run by a wide range of marketing, design, and engineering personnel, these programs must also be verified by means such as those presented in ISO 9001. The need to verify and validate these programs has been demonstrated to the author after personal experience with ISO 9001 audits. Once again, the reader is encouraged to refer to other works on the issue of engineering software management and quality control.[3–5]

At this point, this discussion of the activities of the Design Stage shifts to the topic of Design Verification. Note from the presentation of the ideal ISO 9001 process that the Design Verification activity was considered to be part of the Design Stage. Therefore, some details of the Design Verification activity need to be presented. The basis for this activity is the ISO 9001 standard itself.

As noted in subparagraph 4.4.7 of the ISO 9001 standard, some means of verifying the design (or standard design methods) are specifically mentioned (Figure 5.4). These specific means are performing alternative calculations, comparison with similar proven work, doing tests, as well as reviewing documentation. Each of these means would seem to apply to standard design methods (such as Engineering Standards and computer programs). The review of documentation can be considered as already covered by the approval signature process for the engineering standards. The other means (alternative calculations, comparison with existing work, and doing tests) are now discussed.

Alternative calculations

There should be ample opportunities to make alternative calculations for methods in engineering standards. There are often a variety of methods for calculating the same parameters. For example, a stress calculation (the basic parameter for determining the mechanical strength of a component) can be done using a variety of methods with varying levels of simplification. One can approximate a normal stress by force per unit area, commercial or in-house software can give a variety of solutions, or a book of predeveloped formulas for stress and strain can be used. In each case, the forces or boundary conditions can be approximated in a number of different methods.

ISO 9001 Design Verification for Engineering Standards	
ISO 9001 Paragraph 4.4.7 (ANSI/ASQC Q9001-1994)	**Activity**
"performing alternative calculations"	Calculate parameters based on different analytical techniques (e.g. manual vs. computer)
"comparing the new design with a similar proven design"	Document that technique has been proven effective in previous designs (field tests)
"undertaking tests and demonstrations"	Conduct laboratory tests and compare vs. results of techniques (manual or computer)
"reviewing the design-stage documents before release"	Retain consultants or experts to critique and review techniques

Figure 5.4   ISO 9001 Design Verification for Engineering Standards.

Assuming that one is developing an Engineering Standard for a stress calculation, then one needs to document that these alternative calculations have been used as a check. This can be done in a variety of ways. In one case, the alternative calculations would be included in the standard itself. In this case, the alternative results are presented for a sample calculation along with the results for the method employed by the standard. This has the advantage of being completely documented and controlled. There is not likely to be any question that the design methods have been verified (at least by alternative calculations). This may be particularly useful for a stress calculation since plots or results of specific finite element analysis or FEA models could be made and compared with basic formulas within the engineering standard that covers the stress calculation. The same thinking may easily apply to calculations of weights, centers of gravity, etc. where there are single values to be calculated and where computer programs are readily available.

Verification study

Another way to document verification calculations would be to create an independent document such as a Verification Study. In this case, the standard design method results are presented for a sample problem or problems. The results of alternative methods for these same problems are then worked out and presented for the same problems. The sets of results can then be compared and discussed. In some cases, statistical methods would be applied to consider the results. Finally, a conclusion should be presented by the author of the Verification Study. In this conclusion, the author or expert can state that the design methods are verified by the alternative calculations. Assuming this Verification Study is controlled and is considered a quality record, the design and engineering department should have no difficulty with demonstrated proper use of the Design Verification activity.

Figure 5.5 shows a format for such a Verification Study. The document starts with a Title Page that identifies the engineering standard by number and by revision level. This page also contains the name and signature of the author of the study and the date of issue of the study. Finally, note that a copy number needs to be applied to the Title Page so that the document can be controlled. This document may also require a proprietary statement to declare the information as trade secret. The calculations within such a document should be considered very sensitive; therefore, the distribution of the document should be carefully considered.

Following the Title Page, a section of the Verification Study should be devoted to defining the sample problem or problems. This informa-

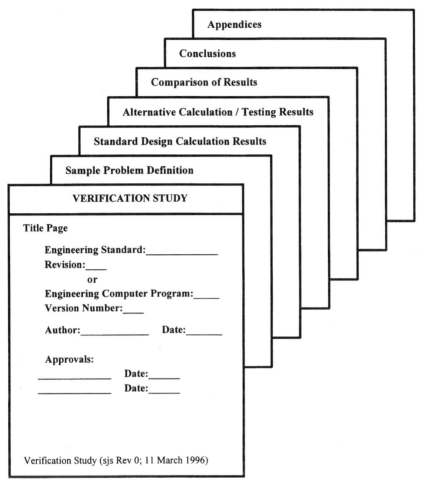

Figure 5.5   Sample Verification Study.

tion is essential to showing that the different calculation methods are applied correctly. The different calculations methods may require different input data, so this problem definition section must present all the data needed by the various methods. Notice that a fictitious or even impractical problem case could be used for comparison. A check on the overall capabilities or robustness of the various methods may be what is desired, and more impractical problems may highlight the differences more clearly.

After the sample problem definition section, there is a section for the standard calculations. This would be the results created by following the design methods presented in the Engineering Standard being verified. The sample problem should be presented on a standard calculation form such as that presented in Figure 5.2. Although every calculation in the method may not have to be presented, as much of the calculation as possible should be shown. In the author's experience, manual calculations are most desirable for these calculations. Even printouts from electronic spreadsheets for Engineering Standards calculations have been a problem with some ISO 9001 audits in the author's experience.

After the standard design calculation section, the alternative calculations can be presented in the next section. This section should also use a standard calculation form, if possible. These calculations should consist of tabular values if the comparison is for particular values (manual calculations versus program calculations). The results can be computer graphics if the comparison is for graphical results (the manual calculations could also be presented as graphs of results). With the computer program results, of course, there should be an indication of the name of the computer program as well as the version number of the computer program. Regardless of whether it is a commercial or in-house program, the program should have a version number assigned. The type and version number for the computer system platform that ran the calculation may also be desirable in the alternative calculation section of the Verification Study.

The next section of the Verification Study would be the comparison of the results between the engineering standard calculation and the alternative calculation. In many cases, this comparison should consist of a table of the two sets of results and difference between them (using statistical methods where possible). Of course, there may be cases where the two calculation methods vary widely; an explanation of the variation should then be presented. Eventually, the author should state in this section whether he or she concludes that the engineering standard calculation is considered verified based on the comparison of the results. A conclusion to the contrary should be considered grounds for removing the Engineering Standard from circulation and archiving the Verification Study; if a calculation method is still required by the organization for that design parameter, a new or better method would have to be found and then a new Verification Study could be prepared for that engineering standard.

Before considering the next means of design verification (comparison with existing work), one should consider the lasting value of documentation such as the Verification Study. These documents would contain a great deal of technical knowledge that is sometimes difficult to capture. This document could include the manual calculations for the design methods and any computer program data (assuming the input data for the program runs are retained).

The manual calculation may be useful in the future for understanding the reasoning behind the formulas presented in an Engineering Standard. The Engineering Standard may present the final formulas to be used, but the manual calculations may reveal information on the derivation of methods and the meaning of the various factors or parameters within the formulas. This may be particularly true for Engineering Standards that are generally run via computer programs.

Comparison with existing work

The next means to be discussed for Design Verification (of design methods such as Engineering Standards) is comparison with existing work. In verifying a specific product that has been designed, this product would be compared with another similar product that has already been released and shown to meet customer requirements. For verifying design methods within an engineering standard, however, the comparison with existing work can be interpreted as demonstrating that proven designs have been created by such Engineering Standards. In other words, if the engineering standards have been used in the past for designs, and these designs have been proven successful, then one can consider the design methods in the engineering standards verified.

The author has successfully applied this interpretation for engineering computer programs. Some analytical computer programs (particularly in large design and engineering organizations) have been used since the dawn of the computer age. Furthermore, these programs have been proven to be effective after many successful designs. In the author's career, there were some computer programs for doing drive train torsional stability that had been used and unchanged for more than 10 years. Hundreds of engineered drive trains had been successfully put in service based on their results. There seemed to be no need to revisit the analytical basis of the methods or the computer programming. In preparation for the ISO 9001 audit for this organization, then, these programs were segregated from more recent programs, and an approval document was created for each program that indicated this past success.

Figure 5.6 shows an example of a Historical Verification form. After clearly identifying the Engineering Standard or computer program (by engineering standard number or computer program name), there is a statement that the results have been proven by a significant amount of successful past applications. Finally, there is an approval signature to demonstrate compliance. Although this technique has proven acceptable in the author's experience, it should not be used as an excuse not to verify even the most proven design methods or com-

---

HISTORICAL VERIFICATION FORM

Engineering Standard:_____
Revision:_____

        or

Engineering Computer Program:_____
Version Number:_____

"The results of the above engineering standard (or computer program) have been used in the past proven designs:

1:_____(release date:_____),
2:_____(release date:_____),
3:_____(release date:_____);

therefore the above standard (or computer program) is considered verified and fit for future application within the terms of its limitations."

Approvals:
_____ **Date:** _____
_____ **Date:** _____
_____ **Date:** _____

Historical Design Verification Form (sjs Rev 0; 11 March 1996)

Figure 5.6 Sample Historical Verification Form.

puter programs. It may be very difficult to produce the original copy of the engineering standard or the computer punch cards from the original author, and therefore it would be difficult to show an auditor that the modern standard or the latest computer program is exactly the same as the original work. Indeed, considering the value of the Verification Study already discussed, it is best to eventually apply the Verification Study approach to all engineering standards and computer programs. The Historical Verification form approach should just be a temporary measure in preparation for an ISO 9001 audit. This is the author's intention for this approach.

Undertaking tests

The last means to be discussed for Design Verification (of design methods) is undertaking tests. These tests are going to differ from tests for a specific product that has been designed (such tests of specific products may be considered prototype testing). In contrast, the tests for verifying design methods may be considered laboratory-type testing. Thus, if there is a specific design method or calculation or algorithm that needs to be verified, then there is an attempt to ascertain if the method is accurate by performing controlled experiments.

The author's approach to this problem has been to formulate a Verification Study again. However, in this case, the comparison is not between two calculation methods (as in the example accompanying the discussion of Figure 5.5 earlier). Instead, the comparison is between the engineering standard's method or computer program's results with the laboratory testing results. Indeed, this is often the best means of verification for the engineering standard or computer program since useful physical and practical data is often collected during well-conceived experiments.

In terms of this Verification Study format, an engineering standard's calculation results could be prepared on standard calculation forms (as shown in connection with Figure 5.2 earlier), and these results can be placed in the usual section called Standard Design Calculation Results. Computer program results could also be prepared using the correct and controlled version of the computer program. The printouts or graphics from the computer program can also be placed in the section called Standard Design Calculation Results (though perhaps renamed to Standard Computer Program Results).

Next the testing results can be compiled. These results would be taken from standard test reports, or copies of standard data recording sheets could be included directly in the Verification Study (at least for tests with a relatively small amount of results and data reduction). In any case, these would be placed in the section of the Verification Study format called Alternative Calculations/Testing Results.

The next section shown in the Verification Study example (Comparison of Results) would discuss and compare the results of the design method (via manual calculation or computer program) and the testing results. This is going to be a difficult process in most cases since the design method or computer program results are not going to exactly match the testing results. Furthermore, the acceptable margin of error or difference between the results is going to have to be established. Assuming these issues can be resolved, the Verification Study which does the comparison of design method results to laboratory testing results should prove a valuable source of information for the design and engineering department. One would then be able to ask designers and engineers for information to back up their methods, and they would be able to refer you to the appropriate Verification Study.

The Conclusions section of the Verification Study format can follow the same approach as alternative calculations in an Engineering Standard. The purpose of this section is simply to state that the computer program or design method is considered verified, and also is fit for use by designers and engineers in the organization.

The use of laboratory testing for verifying design methods introduces the whole subject of inspection and testing in ISO 9001. There is a significant amount of information in ISO 9001 on this subject. However, much of this information is assumed to apply to inspection and testing of a product being designed. Just the same, these sections can be considered to apply to the laboratory type of testing to some degree. The laboratory management of an organization must review the paragraphs of ISO 9001 that apply to "inspection and test" and be certain to comply with the sections that pertain to their specific situation.

Some of the concepts which need to be discussed are standard procedures and calibration, a standard Engineering Test Request, and a standard Engineering Test Report. All these concepts are assumed to apply to the testing activity associated with engineering standards. Note that these concepts constitute another level of demonstrating control over the Design Stage. The philosophy here is that if the product to be covered by ISO 9001 is to meet customer requirements, and the design of that product is based on valid engineering standards, which are assumed to be valid based on results from a testing laboratory, then the processes within that laboratory must have an appropriate level of control.

Perhaps unlike the more fluid and trial-and-error activity of design within an organization, most laboratories within an organization should be accustomed to demonstrating careful control over their activities. This is very much in the nature of laboratories since they must control the testing environment to get useful results. Hopefully, the application of controls for ISO 9001 will not be difficult. The laboratory may already have sufficient written procedures regardless of the existence of ISO 9001.

Engineering test procedures. As usual, the first aspect of control over testing is documented procedures. Just as engineering standards document the design methods and calculations for an organization, engineering test procedures need to document the testing methods for the organization. As before, these procedures must be uniquely identified (perhaps numbered), controlled, and approved by an appropriate expert or authority. Just as there should be an engineering standard for each standard design calculation, there should be a test procedure for each type of test to be performed. These procedures may be developed in a design plan format if the test is only going to be performed once in conjunction with the verification of engineering standards. However, other procedures may be used repeatedly during the course of activities of the laboratory; therefore, they may be kept and refined for future use.

Figure 5.7 shows a sample Engineering Test Procedure. Depending upon the testing being undertaken, this document could be extremely long and complex or it could be simple, perhaps only a single page. As shown in Figure 5.7, the number and name of the procedure should be clearly stated. Furthermore, the number or name could indicate if the procedure is for a single test (such as a single test for verifying an engineering standard). These single-test procedures could be cataloged separately from procedures reused regularly. However, note

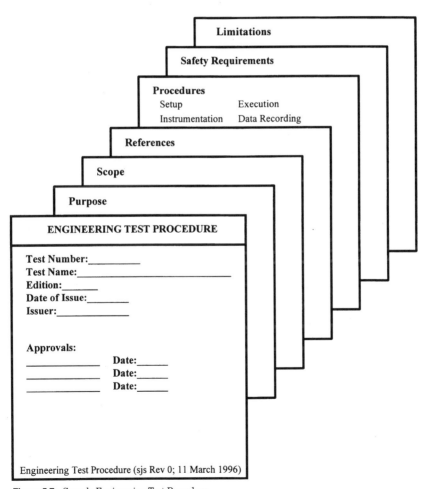

Figure 5.7 Sample Engineering Test Procedure.

that even in an attempt to verify a single engineering standard it is quite possible that the tests need to be run repeatedly to satisfy different problems. In fact, the need to revise and rerun testing procedures is likely to arise when an engineering standard is developed or tested. The results may easily lie outside the expected margin of error, and this outcome may cause the test to be rerun. If there is a standard Engineering Test Procedure on file it should not be difficult to repeat or refine the test procedure.

Following the identification information for the procedure, Figure 5.7 shows that there can be sections devoted to the Purpose, Scope, References, and finally the particular Procedures. As a set of Engineering Standards forms a valuable repository of design methods, so the test procedures can be valuable as well. If a clear Purpose is formulated, then the test procedures can be easily browsed and referenced in the future. After the Purpose, a Scope should be shown to clearly state what situations the test procedure is expected to cover. This information is helpful in the future search for procedures and should assist the expert or manager who must review and approve the procedure.

After the Scope section is a References section. The References are also useful for those requesting tests and for future reference. Many tests can be based on industry standards [such as ASME Power Test Code, Pressure Vessel Code, Society of Automotive Engineers (SAE), ASTM, American Petroleum Institute (API), etc.]. Tests may be based on standard texts published by trade associations for different meters and devices. Applicable sources such as these should be clearly listed in the References section.

After the References section, the sample Engineering Test Procedure shown in Figure 5.7 includes a section for Procedures. This section includes all the steps in the procedure for that specific test. Particularly for the verification tests for engineering standards, the engineer responsible for the design method or engineering standard needs to be involved with the development of the test procedure to be certain that the desired results are achieved. The Procedures for the testing could include instructions for setting up the devices or systems to be tested, installing instruments and transducers, listing the various properties and data to be recorded during the test, and running data acquisition equipment during a test.

As shown in Figure 5.7, a section should also consider safety requirements. The author's knowledge of laboratory testing for design methods has included testing of centrifugal turbomachinery at very high speeds and compressors for volatile gases at high pressures. The safety concerns raised by such tests are enormous, especially considering the experimental aspects of the equipment. If safety concerns are specifically raised by a specific testing procedure, they should be addressed within the Engineering Test Procedure.

The final section of the sample Engineering Test Procedure is for Limitations. As with the engineering standards, there may easily be limitations that affect the validity of the results of the test. If this is a specific concern for a specific test, this concern should be addressed with the Engineering Test Procedure so that future reference to the procedure and the results from the test can be adequately judged for validity.

An approval area is shown on the first page of the sample Engineering Test Procedure shown in Figure 5.7. As these test procedures are likely to be part of the design and engineering department's procedures, the procedures for these tests should be approved as a requirement for Design Control. The Engineering Test Procedure is viewed as a document that plans and directs a specific test to be performed (particularly in support of the Design Verification activity being presented for engineering standards for design methods).

Note that in some cases laboratory testing for verification of engineering standards may involve the design of specialized hardware. This type of hardware could be environmental simulators, testing devices, specialized fixtures or probes, etc. In these cases, drawings may also be created to assist with the design of this equipment. The organization should keep the documentation pertaining to specially designed laboratory equipment as a valuable future reference. An appropriate level of control for drawings from these activities may be similar to the Prerelease type of drawing format presented later in this chapter. It is not appropriate, however, to include the procedures used to design this equipment in the Engineering Test Procedures themselves. The Engineering Test Procedures should be concerned with the use of the specialized equipment, not with its design.

Obviously, some of the Engineering Test Procedures may already exist for a testing laboratory (regardless of ISO 9001), and some procedures used by the facility may be too generic for an Engineering Test Procedure. For instance, data acquisition hardware and software can be configured for all kinds of tests, and the procedures to use this equipment may not have to be part of a specific Engineering Test Procedure for a specific verification test. Instead, a more global procedure can be created and controlled by the management of the testing laboratory for handling this type of generic instruction.

Assuming that there is going to be a set of standard management procedures for the laboratory (in addition to the documented Engineering Test Procedures for specific testing activities), among these standard procedures should be a standard data recording form and a standard test log.

Data recording sheet. Figure 5.8 shows a sample Data Recording Sheet. This is a simple form to be used to record the results of manual tests. This form needs to include clear identification information (such as an engineering test number), date, and name of the person recording the data. Subsequently the form can be used in the recording of manually read data. Obviously, this sheet would not be needed for computerized data recording equipment; however, automatic data recording equipment might be used to make measurements and then values could be transferred from the computer system to the Data Recording Sheet.

Test log. Another essential standard procedure for the test laboratory is a Test Log. A sample Test Log is shown in Figure 5.9 (or at least a single page from a sample log). A *log* is a running document that is continuously updated with pertinent information. In this case, the log is used to record all the events in the progress of the testing. Each day's activities along with the name of the person recording in the log

---

ENGINEERING TEST DATA RECORDING SHEET

Technician: _____   Test Number: _____
Date:_____

Sheet:_____ of _____

Data Recording Sheet (sjs Rev 0; 11 March 1996)   "Trade Secret"

Figure 5.8 Sample Data Recording Sheet.

ENGINEERING TEST LOG		
Test Number: _____		
Test Name:_____		
Date/Time	Technician	Events

Sheet:_____ of _____

Engineering Test Log (sjs Rev 0; 11 March 1996)   "Trade Secret"

Figure 5.9   Sample Engineering Test Log.

should be shown. These activities may include progress and/or problems during the test and data successfully recorded from a test (referencing the proper pages of the Data Recording Sheet that are completed). In some cases, environmental data may also be needed in the Test Log (ambient temperature, barometric pressure, time of day, etc.).

Calibration.   A final type of standard procedure that needs to be developed, maintained, and controlled for the test laboratory is for covering calibration. Indeed, calibration is one of the most important procedures applying to the control of testing of all kinds (including testing to verify engineering standards). These procedures could be documented and controlled in the same manner as the Engineering Test Procedure just presented. However, calibration procedures can also be included in a separate set of controlled calibration documents for easy reference.

Calibration activity is basically an attempt to control the accuracy of the data being recorded from a given instrument or transducer. For instance, a thermometer is only as good as the markings or digital readout shown on it. If the thermometer is not calibrated to accepted standards within a needed level of accuracy, the data recorded from it cannot be trusted. If the data recorded cannot be trusted, then the verification of a design method might not be valid. Thus, it is essential to demonstrate that instruments can be trusted and that they are properly calibrated.

An entire paragraph (4.11) of ISO 9001 is devoted to the control of test equipment. Some of the issues addressed in this paragraph include determining the correct type of instrumentation, inspection and control of the instrumentation, identification of processes for calibration of instrumentation, labeling the test equipment, procedures for handling loss of calibration of instrumentation, and maintaining proper environmental conditions to sustain calibration. A presentation and discussion of all these problems could be quite extensive and is beyond the scope of this book. Most of these activities are more or less common practice for test laboratories employed by industry, and certainly the advent of ISO 9001 is not going be the first time that a laboratory looks at the issue of calibration and keeping instrumentation controlled and within calibration. Paragraph 4.11 should be studied by the management of organizations that employ a test laboratory to be certain that the requirements and their relationship to specific testing is known.

Test software.   Although all the issues relating to calibration of test equipment are not dealt with in this book, there is some interesting notes in this section of ISO 9001. In particular, there is an understanding of the need to control computer software. Test software is specifically mentioned three times in paragraph 4.11. The author assumes that this test software is data acquisition software. This is "real-time" computer programming that samples electronic channels from various and sundry transducers (devices that measure something and convert its measurement into electronic signals). This type of software is very popular and powerful as compared to manual measurement techniques; however, the use of test software probably produces an order of magnitude increase in the opportunities for misapplication of techniques and misinterpretation of results.

Test software may also be responsible for data reduction. After converting the electronic signals into useful units for the test (such as temperature, pressure, force, strain, etc.) based on a calibration conversion, the test software may be responsible for gathering results for many tests and then reducing the data. The test software may also attempt to identify trends and statistically analyze the trends. All these activities have consumed much time and resources in the past, but now the results can be compiled and presented in much less time.

However, just as the use of engineering software and engineering standards can be verified by laboratory testing, you may want to test the test software as well. However, this presents a problem of testing forever. The next step might be testing the firmware that the data acquisition equipment relies upon. Obviously, this situation could continue indefinitely (would you continue to question the validity of the computer language or the compilers or the operating system?), and there would be a diminishing return on the verification of software at more and more detailed levels. In the author's opinion, the test software could simply be verified against manual measurements for a few critical data points. Each testing laboratory management most likely can find an appropriate level of control.

To investigate the appropriate level of control over test software, each of the references to test software found in ISO 9001 is now presented here. In the first of the three specific references to test software in ISO 9001, subparagraph 4.11.1 states that the organization is required to establish and document procedures for controlling test equipment including test software. This is interpreted to mean that there must be standard procedures for the test software just as there are standard procedures such as the Engineering Test Procedure, the Data Record Sheet, the Test Log, and calibration. This should not be difficult to maintain once a set of test software is implemented.

Later in subparagraph 4.11.1, it is stated that test software is required to be checked to prove that it is capable of providing proper results. The author considers this a software validation exercise. That is, just as the Design Validation stage of product design is a test of the product in the customer's environment, the validation of test software involves testing the software in its laboratory environment. To do this, a standard test for the test software should be developed that can be run to show the test software is functioning as required. In the case of test software, the best validation would be versus a noncomputer data recording where possible (some manual measurements and comparisons for some basic data points). To this end, transducer readings from the test software should be performed on benchmark readings.

Subparagraph 4.11.1 of ISO 9001 goes on to state that this "prove-out" activity be repeated at prescribed intervals. Therefore, the validation test that is created needs to be repeatable at regular intervals. In order to satisfy the ISO 9001 audit requirements, these periodic validation checks also need to be recorded, and this documentation needs to be maintained as evidence of control. Regardless of efforts aimed toward ISO 9001 registration, the periodic revalidation of test software is an activity worthy of consideration. As test software gets more sophisticated, the opportunities to draw conclusions without being certain of the integrity of the data are multiplied. To help deal with this complexity, the test software needs to be validated against its benchmark at regular intervals (between series of tests or perhaps daily).

The last reference to test software in paragraph 4.11 of ISO 9001 is in subparagraph 4.11.2. In this case, test software is to be safeguarded. It is to be protected from adjustments which would invalidate the calibration setting and data gathering. This is a whole new area of difficulty for many organizations. With all the computer software used in design and engineering departments (particularly with the advent of personal computers), it has become difficult to control and maintain all the many versions of so many computer programs. How does one know that the test software (or any software in use) is the correct version and that it is the same software that was previously validated? For that matter, consider how many times users have deleted or attempted to alter files on computers without knowing the consequences of these actions. Of course, mainframes and computer networks with different levels of privilege can help protect vital files (such as test software) from being invalidated, but the problem may persist as users determine how to copy, rename, and move files. In some cases, a procedure to control test software may actually be developed in one environment, but new computer environments appear regularly and the procedure may not apply in the new computer environment. This is an issue that each organization needs to address (at least for test software).

Another aspect of safeguarding test software relates to the testing laboratory environment. The computer hardware and software may be adversely affected by the temperatures, humidity, vibrations, etc. that may be encountered in a test environment. To meet this ISO 9001 requirement, the management of the test laboratory needs to consider these factors and be certain that the results from the test software can be protected and trusted for use in the design verification activity for engineering standards.

As stated previously, there may be other requirements in ISO 9001 that need to be reviewed and compared with respect to an organization's use of a testing laboratory. The management of those facilities needs to review each of the test and inspection references in the standard to understand its relationship to a specific testing facility.

At this point, it is assumed that the results from these in-house tests can be approved for use in the Design Verification activity of the Design stage. Furthermore, the results of these tests can be compared to the results of design methods in engineering standards and computer programs using a Verification Study format as presented earlier. If you consider that the laboratory environment is under the proper level of control, a standard procedure for input to the testing process (a test request) and a standard procedure for reporting the output (or results) from the test needs to be presented.

Engineering test request. The engineer responsible for testing for the verification of engineering standards must be careful to make a proper request for the testing. It should come as no surprise that this involves the use of a standard procedure using a standard document or form.

A sample Engineering Test Procedure presented in Figure 5.7 could incorporate this request function (although a separate procedure is probably more desirable). If the Engineering Test Procedure is specifically developed for a test or series of tests to verify a given Engineering Standard, then the procedure document could contain all the information related to the data that is required. However, this information should be segregated into an added Acceptance Criteria section for the sample Engineering Test Procedure. In this new section, each requested data type, location, duration, and measurement uncertainty must be clearly identified. For these tests with a single document covering their plans and activities, the Engineering Test Procedure is more related to the ISO 9001 design plan concept than the standard procedures concept.

As opposed to this single-document concept, it is probably preferable to use the Engineering Test Procedure for only procedures that are to be executed in response to a request for testing. In this case, an engineer requests the test with a separate document (Engineering Test Request). A new Engineering Test Procedure can then be developed as needed to meet requests, or an existing Engineering Test Procedure can simply be used to implement the request. One advantage of separating these activities is the possibility of rejection of the test. In some cases, a test may be requested but then it may be rejected as incomplete, unnecessary, or unaffordable.

The need for a formal procedure for requesting tests can be met by an Engineering Test Request. Figure 5.10 shows a sample Engineering Test Request. In this form or document, an engineer first enters the date and his or her name or title. As mentioned earlier, this document should be in a format that can be rejected or returned to the requestor. Thus, it may be necessary to note the requesting engineer's department or area as well so that it can be returned. After the engineer's personal information, a Test Number is shown. This number probably would not be entered by the requesting engineer. It is assumed that this form is submitted to management, and if it is accepted it would be assigned a number from a Test Request Log.

If we continue to look at Figure 5.10, the sample Engineering Test Request shows that the first section of the document is for the

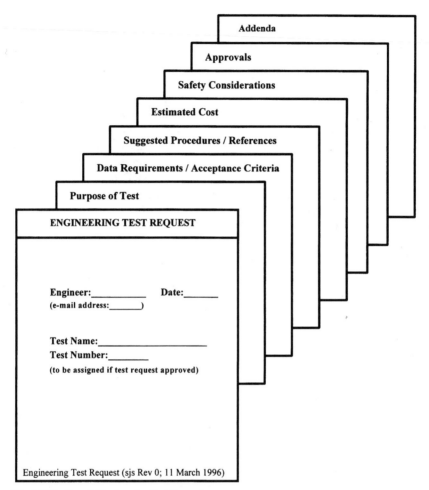

Figure 5.10   Sample Engineering Test Request.

Purpose of the test. This seems obvious but is often overlooked. The real purpose of a test must be known so that the Engineering Test Procedure can be developed to meet that purpose. This is critical since there are often many ways that tests can be designed. There may be a choice of testing in-house versus contracting an outside source; there may be a choice of data collection methods; there may be a wide variety of choices for instrumentation and techniques. In order to make these choices, it is vital to be able to assess the choices with respect to the purpose of the test. In the current case of engineering standard design methods, the purpose would likely state that the test is needed as part of the design verification activity for a given

Engineering Standard or Engineering Standards. Other purposes might include testing a prototype or other new product, or the purpose may be to evaluate a competitive product. These different purposes may lead to the development or planning of different tests.

The next section of the sample Engineering Test Request in Figure 5.10 is the Data Requirements and Acceptance Criteria. In this section, each parameter or type of data required as output from the test must be clearly documented. This can include such information as the parameter to be measured, the location of the measurement, the units of measurement, the duration of the measurement, the statistical sample size, and the uncertainty of the measurement. These parameters and their uncertainties may be considered the acceptance criteria for the test (which must be met or else the test is rejected). There may also be parameters that are actual acceptance criteria along with parameters that are for future reference only or basic interest (i.e., parameters that are not acceptance criteria). The Engineering Test Request clearly shows which parameters are critical to the success of the test and are considered acceptance criteria.

A similarity may be noticed between the Engineering Test Request and the Design Input stage discussed earlier. Indeed, a testing laboratory could consider itself an independent organization that needs to control its processes and requires input and output stage documentation. One of the important concepts in the Design Input Stage is acceptance criteria; the Design Input Stage is to clearly show criteria for accepting or rejecting the design. Following the parallel between the design and development process and the testing process, the engineer requesting a test must clearly show what the criteria are for accepting the results of the test. The testing laboratory can then assess the results of the test with some confidence that the results will be satisfactory to its customer (the requesting engineer). To finish drawing the parallels between the testing laboratory and the entire design and development department, note that the Engineering Test Procedure is similar to the engineering standards of the Design stage, and that the Engineering Test Report presented later in this chapter is similar to Design Output.

The next section of the sample Engineering Test Request in Figure 5.10 is Suggested Procedures. In some situations, the procedures required for a test may already exist or the type of testing may already be commonly performed (regardless of the need for Design Verification activities for engineering standards). In this case, then, the Engineering Test Procedure may already be documented and can be called out in the Suggested Procedures section. The author has seen this situation in various organizations with respect to performance prediction testing. In this case, the testing laboratories already regularly perform performance testing for various needs. Therefore, it may not be necessary to develop new performance testing procedures in order to verify the results of a performance prediction standard or computer program; the procedures may already exist. Thus, the Suggested Procedures section should be used to help prevent the unnecessary development of new Engineering Test Procedures.

The next section of the sample Engineering Test Request is Estimated Cost. This section may or may not be the responsibility of the requesting engineer, but in many cases engineers are familiar with the testing and the resources it is likely to involve. Thus, they should be able to complete this section or obtain the information from the management of the testing laboratory. Note that there is an Approval section in a later section of the Engineering Test Request. As the cost of the requested test is likely to be a factor in that Approval, it should be included in this document.

In conjunction with the Estimated Cost, the requesting engineer needs to document his or her understanding of the safety issues related to the testing. As with the costs, the engineer may have the most current information surrounding the request for testing. Even if the engineer is not the source for this information, the requestor's input should still be documented with respect to safety considerations (in the Safety Considerations section of the sample Engineering Test Request).

The next section of the sample Engineering Test Request is the Approvals section. In this section, the appropriate management personnel need to approve the document. This approval (if successful) signifies that the input to the testing process has been reviewed, and it can signify that the funding can be expended in support of the test. This section can also provide a means of rejecting the request. In order to accommodate the rejection option, space should be available in this section of the document for a reason for rejection. The document can then be returned to the requestor for filing or corrective action.

If the Engineering Test Request is approved, a Test Number needs to be supplied. Perhaps the easiest way to approach the numbering task is to use an Engineering Test Request Log. Figure 5.11 shows a sample log for this function. This log would be maintained by management of the laboratory to track all the current tests and their disposition. This tracking is often needed since tests may be running simultaneously in the same laboratory, and these tests may share resources. The Engineering Test Request Log permits these activities to be tracked by customers, engineers, and technicians. The numbering system could be sequential or it may reflect the type of test or the type of component. The Test Number supplied from this log can then be used in the Data Recording Sheet (Figure 5.8) and the Test Log (Figure 5.9).

ENGINEERING TEST REQUEST LOG

Entry Date	Test Number	Test Name	Reference New Product Code	Estimated Completion Date	Actual Completion Date	Project Engineer	Test Plans	In Process	Complete	Halted

Page: _____

Engineering Test Request Log (sjs Rev 0; 11 March 1996)

Figure 5.11   Sample Engineering Test Request Log.

The use of the input (Engineering Test Request), activity (Engineering Test Procedure), and output (Engineering Test Report presented below) model may be an ideal process for a testing laboratory. However, the nature of engineering testing is that the situation may be more fluid and iterative than indicated by this ideal approach. Particularly when the engineering standard being verified is approaching the state of the art, tradeoffs may need to be made between accuracy versus amount of data collected.

When experimental or testing data is finally collected and analyzed, new factors or problems may be recognized. As the testing is revised to follow better paths to understanding fundamental phenomena, you would not necessarily want to rewrite the Engineering Test Request to amend the changes. In order to avoid a rewrite of the Engineering Test Request document, an Addenda section could be added (as shown in Figure 5.10). In this case, changes to data requirements and acceptance criteria, safety requirements, etc. can be noted. These notes can then be initialled by the requesting engineer and the management personnel which provided the approval for the test.

Engineering test report.   If we move now to the last consequence of the use of testing in support of Design Verification in the Design Stage, the Engineering Test Report needs to be discussed. This procedure corresponds to the output of the testing process. It is a very important concept for the design and engineering department (at least if testing is used as part of Design Verification). In addition, a great deal of important information is going to be stored in the Engineering Test Reports, and they should be carefully maintained and cataloged. Laboratory testing is often an expensive activity; therefore, the results of the testing need to be preserved in order to protect the organization's investment.

The Engineering Test Report needs to document the results of the testing, and it needs to show that the results meet the acceptance criteria. This report also needs to be reviewed and approved. A sample format, such as the one shown in Figure 5.12, should be able to meet these requirements. The first part of the sample Engineering Test Report is a cover page that shows the Test Number, the name of the person preparing the report, and the approval signature or signatures for the report. In addition, the date that the report is prepared should be shown on the cover page.

Depending on the size of the organization and the amount of testing performed in support of the design and engineering department, an abstract and a list of keywords for future searching and sorting

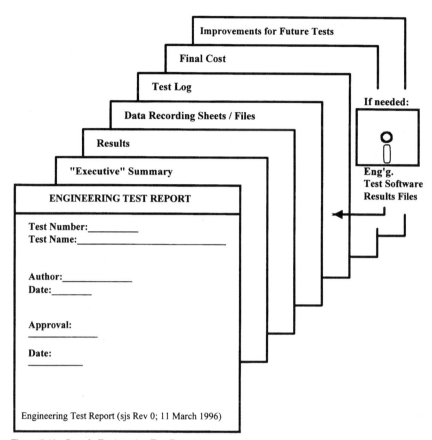

Figure 5.12   Sample Engineering Test Report.

could be included (particularly if a computerized database of the Engineering Test Reports is created and maintained). Following the cover page, there should be an Executive Summary section that provides a means of quickly assessing the overall results. This Executive Summary should at least state whether the results are within the acceptance criteria presented in the Engineering Test Request. This Executive Summary could also be used in a computerized database of the Engineering Test Reports.

The next section of the sample Engineering Test Report is the Results. This section needs to present the results of the testing. Most likely this section consists of reduced or digested results that are presented in some sort of tabular or graphical format. Of course, some testing may only involve a few measurements. In this case, the Results section may simply show the raw data (as a repeat of the

Data Recording Sheet discussed earlier in this chapter). If the Results section presents reduced or statistically analyzed data or trends, then the computer software or the engineering standard used to manipulate the raw data needs to be referenced (if computer software is referenced, the version number of that software needs to be included). The intent of recording this software or method information is that the future reader of the Engineering Test Report should be able to reproduce the information in the Results section by progressing from the raw data in the report or raw data referenced in the report and then using the reduction software or methods referenced in the Results section.

Along with the presentation of the data in the Results section, the acceptance criteria need to be referenced. For each parameter that is specified as an acceptance criteria in the Engineering Test Request, there needs to be shown the actual results of the testing along with the measurement uncertainty. This should permit the report to clearly demonstrate that the acceptance criteria have been met. The intent of recording this acceptance criteria is that the future reader of the report should also be able to verify for himself or herself that the acceptance criteria have been met. Of course, this is also valuable in an ISO 9001 audit situation.

The next section of the sample Engineering Test Report presented in Figure 5.12 is the Data Sheets. This is in keeping with the sample procedures being presented. The standard procedure for collecting data was a Data Recording Sheet. If such a standard document is used (for manual recording of data), then these sheets could be compiled in this section of the report. This section could also be referred to simply as raw data. In the case of computerized data recording, this raw data section could include printouts from the data recording equipment. This raw data section could also reference various computer files; however, these computer files are going to need a whole set of documented procedures for numbering, referencing, retrieving, and preserving the data. Although the preservation of the raw data may not be an ISO 9001 requirement, and it may be difficult to preserve all the information, it is certainly a valuable engineering resource and if possible should be made part of the standard Engineering Test Report procedure.

The next section of the sample Engineering Test Report is the Test Log. Recall that Figure 5.9 presents a sample Test Log for recording the progress of the testing activities (installation, instrumentation, calibration, execution, data reduction, problems encountered, etc.) on a daily or event basis. Although an organization may decide not to preserve all the raw data associated with a test, the daily event information in the Test Log should be preserved to demonstrate control of the testing environment. At the very least, the subparagraph 4.11.2 of ISO 9001 requires that the environmental conditions be properly maintained while the tests are being carried out. The inclusion of the Test Log in the Engineering Test Report should assist in showing that this requirement is being met by the testing facility. As with the raw data, the Test Log information may also prove valuable in future engineering studies relative to the topic covered by the tests. When investigators attempt to draw new trends or conclusions from archived reports, the environmental information may be vital to understanding the validity of these new trends.

The final section of the Engineering Test Report is Final Cost. Recall that the sample Engineering Test Request procedure included a section for Estimated Cost. As mentioned earlier, engineering testing can be an expensive proposition, and in order to track the costs associated with testing the final cost can be included in the Engineering Test Report.

In the Final Cost section, the actual or final costs associated with a test can be presented. This may include just the hardware, instrumentation, customized hardware, and consumables for the test. It might also include labor charges. Often funding for research and development does not include "burden" or "overhead," and it is assessed in a single fiscal year as expense (as opposed to depreciated capital expenditure). However, in the case of testing performed under a subcontract, or when managerial accounting demands it, all the costs including burden and plant expenses may be required in a section such as the Final Cost section.

It is assumed that the Final Cost section is prepared by the person listed on the cover page of the document. However, another department of the organization may be responsible for this information. In this case, that department's records could be referenced or a copy of that department's report could be included in the Engineering Test Report.

With respect to the personnel preparing and approving the Engineering Test Report, some organizations may have technicians prepare the report and then have the requesting engineer approve the report. In other organizations, the engineer may prepare the report based on the data from technicians and then the engineering or laboratory management could approve the report. However, it would be best to avoid having the same person preparing the report preparation and doing the approval. The section of ISO 9001 that applies to the design process states a number of times that information is to be reviewed prior to release; but, the section on inspection and testing just states that it be done according to the organization's design plan or procedures. Throughout this discussion on testing as the means of verification of engineering standards, the author has applied the concepts of the control of the design process to the testing laboratory instead of the control of the inspection and testing process. Therefore, it is assumed that the Engineering Test Report (or the testing output) will be reviewed by someone other than the person preparing the report.

Prototype verification study

Beyond verification of methods in engineering standards, the design verification activity obviously can be applied to the specific design being created. Instead of verifying the methods or techniques within engineering standards, an actual design can be verified. As mentioned earlier, the author considers this activity to be centered around a prototype or other preliminary design that can be verified within the Design Stage. Recall that the Design Validation stage is going to work with the production or final design, so the design verification of an existing product design is assumed to work with a preliminary form of the design (such as a prototype).

The need for verification of a specific design may arise if a design is created without the benefit of a set of engineering standards that dictate the methods to be followed to create the design. Or, the need to do design verification with a prototype may arise if the engineering standards exist, but they have not been verified.

Using design verification directly for a specific preliminary product design or prototype may be considered a more direct application of ISO 9001. Although the use of verification activities for engineering standards is an interpretation of the standard, there is a clear application of the verification activities for prototypes. For instance, the standard clearly indicates that the Design Stage output must meet the Design Input Stage requirements. The Design Input is clearly documented by the design specification discussed earlier, so design verification can be accomplished by simply comparing the Design Input Stage requirements (which are likely to be specific numerical values for specific parameters) with the prototype and verifying that the prototype meets those requirements. However, in doing design verification for the engineering standards there may be no specific design input or target; instead we are trying to develop the best analytical technique for the general benefit of the organization. Therefore there may be no specific Design Stage output comparison with a specific Design Input requirement. Thus, the verification of engineering standards may be more abstract. You should not think, however, their verification is any less important. Considering that the engineering standards, once verified and approved, are going to be applied repeatedly (perhaps without any further study or questions), it may be more important to test engineering standards than to test a given prototype.

As shown in Figure 5.4 for engineering standards, a number of means of design verification are mentioned by ISO 9001. Each of these means can be applied to a specific preliminary product design or prototype as well. Alternative calculations can be used to gain greater confidence in the methods specifically utilized in the design of a prototype. The prototype design can also be compared to an existing, successful design. Also, the prototype itself can be tested or inspected. Finally, the prototype design can simply be reviewed by appropriate personnel.

Since ISO 9001 specifically indicates that design verification measures must be recorded, a standardized test report must be developed for this prototype testing. In addition, meeting minutes must be used for recording the review type of activities. The alternative calculations and comparison with existing design may be more difficult to document.

One possibility for the recording of alternative calculations and comparison with existing designs is a Verification Study. The Verification Study has already been presented in connection with the verification of engineering standards and computer programs (see Figure 5.5). This format may also be feasible for an organization's verification of a prototype design. Another possibility is incorporating this information into the Design File presented in Figure 5.1.

In the interest of using the same format for all kinds of design verification, the Verification Study approach may be considered the preferred approach. Developing a document using this format and carefully controlling a set of Verification Studies can be also be valuable in an ISO 9001 audit situation. In this case, all the information is centralized.

Figure 5.13 presents a sample Verification Study applied to the prototype design for alternative calculations and/or comparison with a successful existing design. The cover page for this document does not show the engineering standard or computer program to be verified as in the previous Verification Study. Instead, this document shows the new product identification (for instance, the new product code taken from the Development Project Master Log presented in Chapter 4). The Verification Study should clearly indicate which design is being verified. As the document may be refined during the life of a development project, a Revision number is used along with the name of the author, the date the document is completed, and space for an approval signature (since design stage documents are to be reviewed). Since this document is going to be controlled for the use in an audit situation, there is a copy number shown.

Figure 5.13 Sample Prototype Verification Study.

After the title page information, the sample Verification Study for a prototype shown in Figure 5.13 has separate sections for the prototype design information and the verification information. The prototype design information may include sketches, drawings, and standard calculation forms. This section should at least contain a written description of the prototype design. This information is vital for the reviewer or reader to assess the validity of the verification information.

After the documentation for the design, a section is used for Alternative Calculations or comparison with existing, successful designs (or both these sections could be present). The author of the Verification Study needs to present the verification information. The alternative calculations can be presented on standard calculation forms. A comparison with existing design section can show information about the similar proven design using data from field testing and/or service records.

Finally, the sample Verification Study for prototypes contains a Conclusions section. This section should compare the alternative calculations with the prototype design calculations. This section should compare (feature by feature, perhaps) the similar design and the prototype design. These factors should be compared and contrasted, and then the author should express whether the design can be considered verified based on the comparison (or whether the design is likely to meet the Design Input requirements). Assuming that the author feels the design is verified by the means presented in the Verification Study, the reviewer can agree to the assessment by signing the Approvals section on the cover page.

Prototype testing

The next means to be discussed of design verification for a specific preliminary product design or prototype is testing. This may be the most common and direct form of design verification. However, as with the other means, there must be a means of documenting or recording the design verification results. The obvious approach in this case is to develop standard prototype testing documentation similar to the engineering testing documentation discussed earlier (Engineering Test Request, Engineering Test Procedure, and Engineering Test Report).

For the Engineering Test Request (Figure 5.10), there may be some changes required for use with a prototype test. These changes could be incorporated into a unified Engineering Test Request form, or a separate Prototype Test Request form could be created. The differences between these formats may be driven by the product and the regularity of the testing. If the product designed by the organization is constantly being tested, then the Estimated Cost and Safety Considerations may not need to be completed by the requestor for each test. Instead there may be standard burden charges and safety standards listed for these commonly performed tests. This is in contrast to the testing for verifying engineering standards where the testing is likely not to be regularly repeated once the engineering standards are successfully verified.

A Prototype Test Request must be uniquely identified (by a Test Number, for instance). This Test Number should be unique for a particular development project and should be different than the number used for the engineering tests. However, as with the engineering tests, the number could be entered and maintained in a unified Test Request Log (as shown in Figure 5.11).

A Prototype Test Request should include the remaining sections shown in Figure 5.10 in the same manner. The Purpose should still be stated, to be certain that future referrals to the document clearly indicate the kind of test, what model has been tested, etc. The Data Requirements and Acceptance Criteria section would still be essential in the Prototype Test Request. As Design Verification for the product design may be relying on the results of this test, the acceptance criteria of the design (as defined in the Design Input Stage) should be available and stated in this section. There may also still be a Suggested Procedures section. However, since the testing may be more standard than specialized (as in the engineering standards testing), the procedure may not have to be suggested as often. Instead, the testing procedure may be standardized and the Prototype Test Request may dictate the test procedure to be used in the testing.

The sample Engineering Test Procedure shown in Figure 5.7 can probably be used directly as the format for a sample Prototype Test Procedure. The prototype testing procedure should still be clearly identified by a number and name. The prototype testing procedure should still have a Purpose for guiding the reader in assessing the applicability of the test in new situations. The prototype testing procedure can certainly still have a Scope (indicating what type of product or what product line is covered by the procedure, for instance). The prototype testing could still have a References section since there may be applicable industry standards for various kinds of testing (SAE, for instance). The Procedures section can still include subsections for test setup, instrumentation, and data recording. Following the Safety Requirements section, there must be an Approval section to show that the procedure has been reviewed, and this Approval section should provide for a prototype test request to be rejected and the reasons for the rejection, stated for the benefit of the requesting engineer.

As shown by the samples in Figures 5.8 and 5.9, there should still be standard forms for data recording (if manually recorded) and a Test Log for recording the events that occur in the testing activity. In this case, however, the forms would show a test number for prototypes instead of a test number for engineering standards.

A sample Prototype Test Report may also be based on the sample Engineering Test Report shown in Figure 5.12. The title page would then show the Test Number, the date the report is completed, the signature of the person preparing the document, and the signature of the person reviewing and approving the document. As before, the Executive Summary should at least state whether the prototype test results meet the acceptance criteria. The Results section would need to detail the data collected in the prototype testing (as in tables, graphs, charts, etc.). The Data Sheets or the computer file information needs to be presented in the Results section (or as an appendix) so that a future reader of the document can study the testing. The standard Test Log needs to be included as well. The Final Cost section may also be included. As with the Estimated Cost, this may depend on the regularity of the testing or a specific organizations handling of charges related to testing.

Prototype inspection

One aspect of prototype testing that may differ from the engineering standards testing is inspection of the preliminary product or auditing. Since the prototype testing involves testing on a specific product design, and since the Design Input Stage produces a clear list of criteria for that design, one can consider creating a test procedure which simply documents how someone examines the product from the point of view of the acceptance criteria. This test procedure could then specify who is to inspect the prototype, how the prototype is to be inspected, and how to handle nonconformance (where the inspector finds a problem that must be corrected). In order to request the inspection, the Prototype Test Request could simply call for the inspection procedure by name or number. A Test Number from the Engineering Test Request Log could be used in the same manner as before (see Figure 5.11).

In the prototype inspection or audit procedure, someone familiar with the product design and the design input requirements would inspect or audit the preliminary product. In this case, the standard Data Recording Sheet presented earlier may not be appropriate. Instead, items that are considered problems would need to be documented and studied (as opposed to simply writing down measurement values from instruments). Unlike the Data Recording Sheet, there also needs to be an accompanying process for handling the agreed-upon problems with the design. The areas of noncompliance could be related to performance, appearance, safety, etc. Furthermore, instead of using Data Recording Sheets in the Engineering Test Report, a prototype testing report can include Inspection Problem Incident Sheets. If no problems are discovered in the inspection activity, there would be no such sheets in the Engineering Test Report for the prototype test and the Results section could indicate that no problems were discovered.

Figure 5.14 shows a sample Inspection Problem Incident Sheet. This sheet would be standardized and its format specified by the Engineering Test Procedure for this inspection process. The sample format shows the usual identification information such as a Test Number and the inspector's or technician's name. Following the identification information, a table for entering problem items is shown. These problem items are areas or aspects or behaviors of the preliminary design which have failed documented criteria (for performance

INSPECTION PROBLEM INCIDENT SHEET				
Test Number: _____ Inspector Name:_____				
Date/Time	Item Number	Problem Description	Failed Criterion	Confirmation
			Sheet:_____ of _____	
Inspection Problem Incident Sheet (sjs Rev 0; 11 March 1996) "Trade Secret"				

Figure 5.14   Sample Inspection Problem Incident Sheet.

or safety, for instance). The inspection may take more than one day, so the date is shown first for each problem item. The next column in the table is for a problem Item Number. Each item discovered by an inspector should be uniquely identified for tracking of the items. Of course, if more than one technician or inspector is doing this activity, then the Item Number would have to indicate the inspector's name. The different Problem Incident Sheets can then be compiled, and each problem item would still be uniquely identified for confirmation. The next column in the table is the Problem's description. In addition to clearly identifying the problem discovered in the inspection activity (location, behavior, etc.), the criteria that the inspector feels has been violated can be shown.

After the column for the Problem Item description and criterion, Figure 5.14 shows a column for Confirmation. This column shows the results when the problems that have been discovered are reviewed, and is intended to indicate that the problem in fact needs to be tracked and corrected.

Once items are discovered and confirmed, the confirmed problems need to be tracked. This means listing the problems and showing whether corrective action has been undertaken or completed. Figure 5.15 shows a sample format for a Problem Incident Log. Each confirmed Problem Item is listed by number. These items are taken from the Problem Incident Sheets. After the Item Number, the corrective action for the problem is entered in this log.

After the Corrective Action Required is shown in the format shown in Figure 5.15, the Person Responsible for the corrective action is listed. Subsequently, a column [Closed (Confirmed)] is shown in the Problem Incident Log for indicating whether the problem item is closed (i.e., the corrective action has been taken). It should be assumed that the Corrective Action Required will be implemented before the Design Stage is considered complete.

Note that the inspection procedure for prototype testing could be computerized or automated. For example, the inspector could enter the problem items into an electronic database instead of writing them down on the sample Problem Incident Sheet. The list of items could

PROBLEM INCIDENT LOG			
Test Number: _____			
Item Number	Corrective Action Required	Person Responsible	Closed (Confirmed)
		Sheet:____ of ____	
Problem Incident Log (sjs Rev 0; 11 March 1996)		"Trade Secret"	

Figure 5.15   Sample Problem Incident Log.

easily be printed using this database. The confirmation of the discovered problems could be another entry in the database made by the person responsible for the test or prototype. The Problem Incident Log could then be automatically generated by the database software. The open and closed (confirmed) items could be printed whenever needed.

One problem with the electronic database system is controlling access to the database. This might be accomplished, however, by passwords associated with the database software or with a computer network. In a computer network which uses passwords and varying levels of privilege to data or programs (such as read/only, read/write, etc.), access to the database could be restricted for showing the problem items that are closed. Of course, this approach may require the institution of controlled procedures relative to the maintenance of the computer systems and software.

Whether a paper-based or computerized system is used, the prototype inspection and testing needs to be controlled and documented. This type of testing is designed to meet the requirements of design verification by checking an output from the Design Stage (a prototype or preproduction model) versus the input to the Design Stage (design specification). This type of prototype testing might be used alone to provide design verification (probably for a single unique development project) or it might be used to supplement the laboratory type of testing for Engineering Standards (particularly if many new designs are created with engineered variations to a basic design). Finally, a design may be verified by the comparison of the new design with existing successful designs.

Design Review in the Design Stage

The overall ideal ISO 9001 product design and development process presented in Chapter 3 indicated that a design review activity must be included in the Design Stage. This design review is assumed to be required by ISO 9001 for design verification of a specific product design (similar to the prototype testing activity). The design verification for a specific product design should concentrate on determining whether the design meets the requirements or specifications created in the Design Input Stage. For the specific product being designed, this Design Stage Design Review needs to compare the Design Input Stage documentation (design specification) with the current state of the design. Although ISO 9001 indicates that design review meetings can be held at different, appropriate stages of design, the design review being presented is assumed to occur at the end of the Design Stage. If it is successfully completed, the Design Stage is considered at its end and the Design Output Stage can begin.

As a reminder, recall that the Design Output Stage (for this ideal process) also has a design review meeting. However, the Design Stage Design Review is supposed to differ from the design review activity in Design Output Stage (which is described in detail in Chapter 6). The design review within the Design Output Stage has the benefit of having all the documentation approved and ready for release (drawings, bills of material, installation instructions, part lists, etc.). The design review in the Design Stage, however, may only have the benefit of documentation such as sketches, testing results, and Verification Studies.

The author's development of the ideal ISO 9001 process using two different design reviews is based on the assumption that the design project is large enough to require different personnel for the basic product design versus the product documentation. In the author's experience, there may be a core group or team of people most involved with the product's basic design. These people actually go through the iterative process of attempting a configuration, testing it, and then reworking the configuration (as discussed at the beginning of this chapter). There may also be a separate group or team for preparing the package for sending design documentation to the production groups. The core people, however, are considered best suited for looking at the technical capabilities of the design and comparing it with the technical requirements of the design specification (i.e., they are the best suited for the design review activity in the Design Stage).

Another advantage to having the design review at this point in the process is handling changes needed for the design. The people involved with the basic product design and its results from testing are also best suited for determining the solution to problems or inadequacies found relative to the design specification. These people may also need to decide whether the design specification must be changed since a new conceptual design may never be able to meet the originally desired specifications. In the author's opinion, all these factors are different at the end of the Design Output Stage where the entire documentation package needs to be verified for completeness and adequacy.

Implementing the design review activity for the Design Stage should not be difficult. The design review activity is assumed to take place at a meeting. ISO 9001 requires that the meeting be documented and that the documentation be controlled, as usual. The documentation could be minutes from the meeting, or it could be a standard form that is filled out and signed.

The Design Review subparagraph in ISO 9001 specifically mentions that participants at the design review represent all functions concerned with the design stage as well as specialists. Therefore, one must give some thought to the list of participants in the design review. These participants would have to be informed of the meeting and its agenda, so the first step in initiating a standard format for design review meetings would be to publish a Design Review Notice informing people of the upcoming meeting. Figure 5.16 shows a sample Design Review Notice. The form would probably be distributed to the participants after an acceptable date, time and place is established.

The form in Figure 5.16 shows a New Product Code so that the participants know which design project is going to be discussed. In some cases, the participants may need to study the status or results of the project prior to attending the meeting. After the product code, a project engineer or other responsible person should be specified. This person could be the one calling the meeting and chairing it.

Finally, the participants could be listed by department. Some of the departments to consider are the design or engineering department, the department responsible for testing (if used in the design verification activity), the product safety department, a human factors depart-

---

DESIGN REVIEW NOTICE   ☐ Design Stage
                             ☐ Design Output Stage ("Final Review")

New Product Code:_____

Project Engineer:_____   (e-mail address:_____)

Meeting Date:_____   **Meeting Time:**_____   **Meeting Location:**_____

Participants:

                                                        Present:

    1. **Design:**_____   ☐

    2. **Testing:**_____   ☐

    3. **Product Safety:**_____   ☐

    4. **Human Factors:**_____   ☐

    5. **DFM:**_____   ☐
       (Design For Manufacturing)
    6. **Other:**_____   ☐

Design Review Notice (sjs Rev 0; 11 March 1996)

Figure 5.16   Sample Design Review Notice.

ment (if it is not part of the design department), and the Design For Manufacture (DFM) specialist or department.

DFM is particularly important in the Design Stage Design Review. Although product designers are going to be attempting to create manufacturable designs, there are clearly benefits to bringing in specialists to review the design for the manufacturing issues (particularly with new concept designs that may need new manufacturing methods). If there is no formal DFM group in the organization, its function could be filled by the Manufacturing or Industrial Engineering department. Better yet, the Design Stage Design Review could invite workers from manufacturing (or shop floor) to attend. If one is lucky enough to have workers that can communicate effectively to designers and engineers, many useful ideas may be brought to the discussions. Although this may not be an ISO 9001 requirement, an ISO 9001 registration effort may be an effective means for pushing this inclusion of different departments into the design process.

Of course, the basic task of the Design Stage Design Review meeting is to verify that the design meets the requirements of the design specification. This basic aim may or may not relate to actual production, particularly if the performance and the basic feasibility of a new conceptual design is being addressed. Of course, in most cases even the basic conceptual design has repercussions for manufacturing so it may always be better to include these representatives.

The sample Design Review Notice form shown in Figure 5.16 shows a column after the names under the heading Present. This form could be used during the meeting itself to record the attendance of the designated participants. This form records whether the proper representatives attended the meeting.

If necessary, an agenda could be attached to the form. If the development project is large enough, multiple meetings might be held for different assemblies or subassemblies of the product. In this case, the form shown in Figure 5.16 might include another item to fill out to indicate what part of the product is going to be reviewed at the meeting.

The content of the design review meeting might be standardized for a given organization. Figure 5.17 shows a standard form or checklist for this meeting. After the project identification information (new product code, project engineer, date of the meeting, etc.) there is a list of different aspects of the design to be reviewed. This list includes the acceptance criteria and the regulatory requirements. This information should be listed in the design specification for the project, and can be copied or referenced on this form. As the representatives at the meeting agree that each criterion or requirement is met, then the item should be checked off or "ticked."

```
┌─────────────────────────────────────────────────────────────────┐
│                  DESIGN STAGE DESIGN REVIEW                       │
├─────────────────────────────────────────────────────────────────┤
│       New Product Code:_____              Date:_____         │
│       Project Engineer:  _____  (e-mail address:_____)         │
├─────────────────────────────────────────────────────────────────┤
│       Acceptance Criteria:                   Check if met:       │
│           1:_____       ☐              │
│           2:_____       ☐              │
│           3:_____       ☐              │
│                                                                   │
│       Regulatory Requirements:                                    │
│           1:_____       ☐              │
│           2:_____       ☐              │
│           3:_____       ☐              │
├─────────────────────────────────────────────────────────────────┤
│                        Check if done:                            │
│       Safety Review:          ☐                                  │
│                                                                   │
│       FMEA:                   ☐                                  │
│                                                                   │
│       DFM Review:             ☐                                  │
│                                                                   │
│       Maintenance Review:     ☐                                  │
├─────────────────────────────────────────────────────────────────┤
│       Approvals:                                                  │
│       _____  Date: _____                               │
│       _____  Date: _____                               │
│       _____  Date: _____                               │
│       Design Stage Design Review (sjs Rev 0; 11 March 1996)      │
└─────────────────────────────────────────────────────────────────┘
```

Figure 5.17   Sample Design Stage Design Review.

The next aspect of the design review meeting that is listed in the sample form in Figure 5.17 is safety requirements. In this stage of the design and development process, the design should be considered changeable. Since the core design team is involved in this review, and the formal documentation of the Design Output Stage has not yet started, it should be possible to change the design for safety (or other) reasons.

You should note that this design review activity is assumed to be doing more than just looking at the design specification and checking that each criterion is met. For the safety requirements, for example, the representatives at the design review must be proactive. They need to imagine "what if" scenarios with the design. Then they should consider the effects of these scenarios, and whether the design needs to be changed to prevent injury or damage or loss. Theoretically it should be much easier to alter the basic design at this Design Stage than to wait for the end of the Design Output Stage.

Furthermore, at this point in the process, the design could be experimental or a new concept of some kind. Therefore, the consequences of using the new concept in the market might not be known. Those most intimately involved with the design (those attending the design review) need to consider the environmental consequences, the social consequences, the national security consequences, etc. of the design. This may even be seen as an ISO 9001 requirement, since these other consequences could be broadly interpreted as safety requirements (if "safety" is interpreted as dealing with the environment or society or national security as opposed to simply a customer using or buying the product).

As the safety review activity (in either the broad or narrow interpretation) is completed, the safety review box shown on the sample form can be checked off (assuming that there were no problems that prevent the design from being approved). If there is a safety problem raised that cannot be approved, then corrective action is going to have to be taken. The corrective action may be to alter the design or perhaps further testing to be certain of some issues with respect to the design.

Another aspect of the design review which may be added and that is similar to the safety review would be failure mode analysis. This activity would not necessarily concentrate on the injury or loss effects of the design, but the effects of failure on the functioning of the design. In this case, the representatives at the design review should be proactive with respect to what may happen to the functionality of the design when different parts of the design fail at different times or in different environments. As with the safety review, hopefully the design can be easily changed to handle problems brought to light in this aspect of the design review. And, as with the safety review, the sample form in Figure 5.17 could simply be checked off at the failure mode analysis item in the list. In some cases, a failure mode specialist may be required to study the design prior to the meeting so that his or her findings are already available at the meeting.

The final aspect of the design review that is shown in the sample form in Figure 5.17 is DFM. This is another area where the representatives at the design review meeting should be proactive. The design specification is probably not going to provide a numerical target for a design parameter that can be measured for manufacturability. Instead, the representatives at the design review must imagine the advantages or disadvantages of the design with respect to ease of manufacture and/or assembly. As this aspect of the design review is completed and the representatives at the meeting are satisfied with the design, this item in the list can be checked as well.

Note that other aspects could be added to the design review. Some of these areas might be ease of operation, ease of maintenance, human factors engineering, and training. Although these areas may not be specifically called for in the ISO 9001 standard, the design and engineering department should make use of the ISO 9001 requirement for design review meetings to its best advantage. If an organization has been placing too much emphasis on completing a design as fast as possible, as opposed to taking the time to be certain it is right, an expanded design review meeting may be called for as part of an effort to attain ISO 9001 registration.

The final section of the sample design review format in Figure 5.17 is for approvals. The person chairing the design review meeting should check each of the items in the checklist on the form. At the end of the meeting (or meetings) all the representatives can sign the form to indicate that this design review is complete. As this form is then complete, it needs to be retained and controlled for use in future audits as a quality record that the design review requirement as part of the design verification activity has been completed. This form might be kept in a separate file of all these forms for the projects or it may be kept in the larger project file that is discussed as part of the Design Output Stage in Chapter 6.

The design review meeting is the last topic presented with respect to the design verification activity of the Design Stage. To reiterate, the author considers design verification one of the most important issues for designers and engineers involved with ISO 9001. Design verification may easily take the most time and effort in preparing for the first ISO 9001 audit. As this effort progresses, each design and engineering department should carefully study their processes and take the opportunity to institute meaningful verification methods.

Graphics

One practical consideration of the Design Stage that needs to be considered as part of an ISO 9001 effort is the use of graphics. It is the Design Stage which determines the configuration of a product design, and many issues concerning configuration are geometric. That is, the Design Stage is going to determine what the product looks like. The product's geometry and materials are critical to the success of the product or its ability to satisfy the customer. This information is found in graphical representations; therefore, these representations become a quality issue. Thus, the use of graphics in the Design stage needs to be a consideration in the development of standard procedures for satisfying ISO 9001 (at least under its heading of Design Control).

Note that graphics in the Design Stage may include sketches, drawings, physical models, and computer-generated three-dimensional models. In a two-dimensional environment, designers or engineers would create a simple graphical representation of a new product or part. This is generally called a *sketch*. Based on sketches, drawings would be created (perhaps for use in prototype construction). Often the drawings usually are more formal graphical representations than the sketches.

Sketches

In order to satisfy ISO 9001 for Design Control, the sketches should be drawn on a standard form (particularly if they are to be included in an engineer's Design File). Figure 5.18 shows a sample Engineering Sketch form. This form needs to show the engineer's or designer's name as the person creating the sketch. The form also needs to show the development project identification (shown as a New Product Code). Next a date for the sketch is shown, and then there is space for the sketch. The last item shown on the sample Engineering

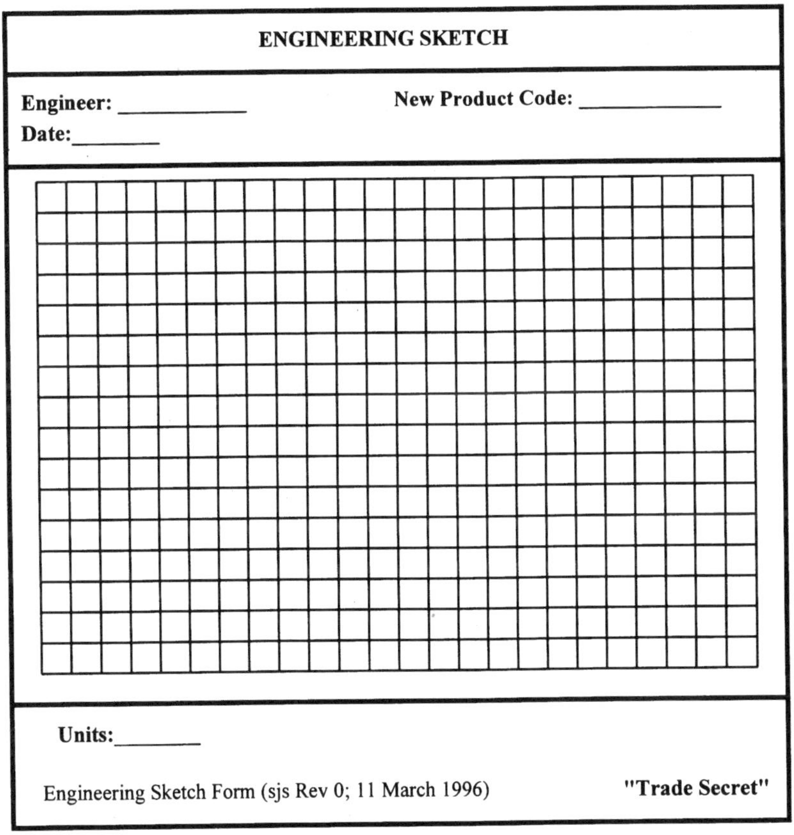

ENGINEERING SKETCH

Engineer: _____   New Product Code: _____
Date:_____

Units:_____

Engineering Sketch Form (sjs Rev 0; 11 March 1996)   **"Trade Secret"**

Figure 5.18 Sample Engineering Sketch Form.

Sketch form is Units. This should be used to indicate the unit system used for the dimensions and parameters shown in the sketch (such as millimeters or inches and at what scale).

Obviously, many designers and engineers are going to use a computer system for graphics such as the sketch. This class of software is referred to as computer-aided drafting (CAD) in this book, and it implies a system that provides computer graphics only. Computer software that assists the designer or engineer by making design calculations is called engineering software or computer-aided engineering (CAE) software in this book.

If a CAD system is used for sketches, a standard drawing format for the sketches needs to be created and made available to engineers and designers. This drawing format needs to include the same basic information shown in the Engineering Sketch form; but in this case, the information is placed on the border of the computer graphics space in something referred to as a drawing format (refer to Figure 5.19). In the sample Engineering Sketch Format shown, the pertinent information is Date, Drawn By (if a draftsperson assists the designer or engineer), and Engineer shown in the lower left corner, and New Product Code, Part Name, and Units in the lower right corner.

One problem with using CAD for sketches is that the computer graphics output might be mistaken for formal drawings (even though the format does not include all the information that is found on a formal drawing). This can be due to the fact that the computer-generated output may be as clean and presentable as the formal or released drawing from the Design Output Stage. To prevent this situation, the Engineering Sketch Format needs to include a clear PRELIMINARY note (as shown in Figure 5.19).

The sketches (in the CAD or manual format) are likely to be eventually used in the creation of drawings. Drawings are more formal graphical representations and are considered more formal documentation than the sketches. Drawings, in most cases, form the foundation of the configuration of a product design as it moves beyond the Design Stage.

In the author's view of the ideal ISO 9001 process, a separate Design Output Stage is considered to handle the development and creation of the drawings (and other documentation) for production or manufacturing of the product. However, the Design Output Stage may be seen as arbitrary timing since there may be drawings prepared in the Design Stage as part of testing and prototype activities.

Although the boundary between the Design Stage and the Design Output Stage may be difficult to determine, there are advantages to making the distinction between these activities. The Design Output Stage is going to be centered on preparing all the documentation for

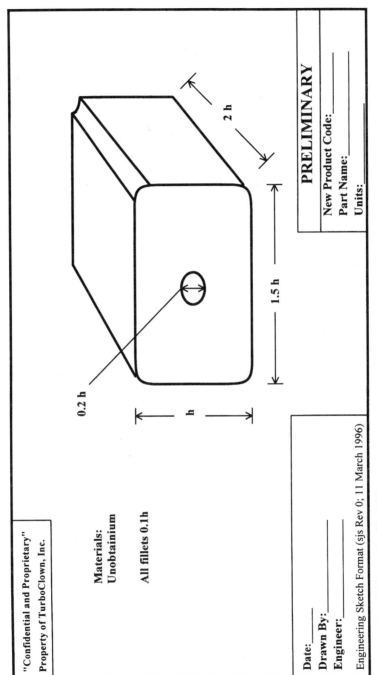

"Confidential and Proprietary"
Property of TurboClown, Inc.

Materials:
Unobtainium

All fillets 0.1h

0.2 h

2 h

1.5 h

h

PRELIMINARY

New Product Code: _____
Part Name: _____
Units: _____

Date: _____
Drawn By: _____
Engineer: _____

Engineering Sketch Format (sjs Rev 0; 11 March 1996)

Figure 5.19 Sample Engineering Sketch Format.

release to the production environment with all necessary details. By contrast, the Design Stage is going to be centered on just determining the configuration of the design.

If the final documentation is being prepared while the basic design is still changing, then there may be a great deal of rework. If the drawings prepared in the Design Stage have to be detailed completely (with all dimensions, references, tolerances, part numbers, formats, and approvals), and then some basic geometric parameters of the design have to be changed after prototype testing, then all these drawings may have to be revised. There also may be lost effort in renumbering the parts and reapproving the changes.

In addition to rework of drawings, the Design Stage activities may be hampered by the requirement of preparation of completely formal drawings. Once these drawings are created, there may be a reluctance to make fundamental changes in the design. Furthermore, specialists involved with the basic design may be used unproductively in the detailing activities. Of course, parametric design software and automated detailing may remove some of these difficulties.

Prerelease drawings

Although the creation of formal drawings in the Design Stage may have disadvantages, there may still be a need for more formal graphical documentation than sketches during the Design Stage. Also, one needs to avoid the recreation of the drawings in the Design Output Stage without making use of the Design Stage graphics. One possible solution is the creation of a standard procedure for Design Stage drawings. This standard procedure would create drawings separately from the Design Output Stage drawings. These drawings could be considered prerelease drawings (as opposed to preliminary sketches).

In addition to being different from final drawings in appearance, prerelease documentation could use preliminary or prototype part numbers or identities. Use of this technique is not necessarily an ISO 9001 issue (as long as a proper level of control over the design activity is maintained). As long as the procedure is documented and followed there should not be a problem for the ISO 9001 effort.

In this situation of using a prerelease product configuration, the organization needs to determine the best tradeoff between the flexibility to easily change the design (its graphical representation, part numbering, bills, etc.) versus the risk of not maintaining a completely disciplined database of the design configuration. Of course, all the work of creating a completely consistent set of documentation must be done eventually, and we may assume that it is best to maintain all the controls possible during the Design Stage. However, the author views the Design Stage as the most important activity within the entire organization and stresses that innovation here must not be hampered unless absolutely necessary.

Although the product configuration is being determined in the Design Stage, and the drawings may not be in the final form for use in the production environment, the design must still carefully consider manufacturability. Just because these drawings are not going to be released directly to a Production Control department, this should be no excuse for creating a design without regard to the production environment. Indeed, the relaxation of excess controls in the Design Stage should enable designers to give more thought to manufacturability (or related issues), not less.

Assuming that the prerelease drawings approach is taken, the use of drawings in the Design Stage need to be standardized. These prerelease drawings could be issued, revised, and controlled in a separate environment from the final or released drawings, but they must be easily identified as prerelease. Figure 5.20 shows a format for this type of drawing. This format shows more information than the sketch format in Figure 5.19. This format includes formal information such as a drawing Title, a Part Number (considered a prerelease part number), and a Checked By (for indicating the drawing is checked for proper drawing technique).

Furthermore, the drawing may be sent to a prototype shop or a vendor for manufacture. There is a part number indicator, but as mentioned earlier, this may be a preliminary or specially identified part number (which is going to have to be mapped to an actual controlled part number at some point).

Note that the drawing format has a clear PRERELEASE indicator. This must be understood by the receiver of drawing to mean the drawing is controlled in a separate environment, and that only Design Stage activities such as testing and prototype manufacture are to be performed based on those drawings.

The use of prerelease drawings during the Design stage should not produce difficulties with creating drawings in the Design Output Stage. Since most drawings are going to be done on a CAD system, the prerelease drawings can simply be loaded into the final drawing database and the format updated to the proper Design Output format (which would not have a PRERELEASE note). Of course, if prototype-based part numbers are used, then the numbers shown in the drawings would have to be updated to the final numbering system.

The use of prerelease drawings during the Design Stage should not produce difficulties when drawings need to be revised. Indeed, there should be less confusion in dealing with drawing versions or "revision levels" during the Design Stage if a prerelease format is used. If final Design Output Stage drawings are demanded in the Design Stage, and some drawings need to be revised in the Design Stage due to test-

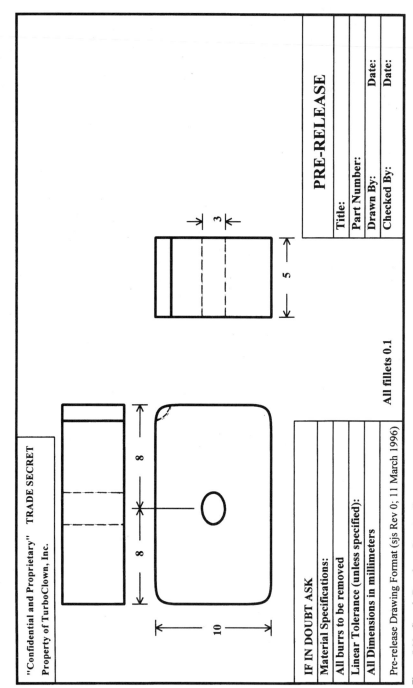

Figure 5.20 Sample Prerelease Drawing Format.

ing or other considerations, then there will be a variety of revision levels going into the production environment for the final drawings. However, if the prerelease drawing method is used, there can be revision levels as needed for the prerelease drawings during the Design Stage. When the Design Output Stage is entered, all the final drawings can be set to a consistent revision level (such as 0, 1, or A). When a drawing has a revision level above the starting value, this is a clear indication that a drawing has been revised since release (as opposed to allowing the possibility of changes having been made during the preliminary design activity).

Thus, allowing the prerelease drawing method may seem to reduce excessive control in the Design Stage. It may also be valuable in maintaining control over the design in later aspects of the product life cycle. Of course, as mentioned earlier, this activity assumes the ideal ISO 9001 process, and each organization needs to determine whether this method is really desirable.

One activity associated with drawings that should be included in the Design Stage drawings whenever possible is tolerances (the allowed variation in dimensions of parts and characteristics of parts). Consideration of special tolerances in parts requiring nonstandard values is going to be based on the basic configuration of the design. This consideration should be handled as closely to the basic design activities as possible. Although an organization may have a standard or basic tolerance for manufactured parts, the Design Output Stage personnel may not be qualified to determine which parts should not use the standard tolerance.

Solid models

Although two-dimensional drawings with standard views can provide clear understanding of the configuration of a product or part, the use of two-dimensional drawings can be a source of ambiguity when the drafting person makes errors. The advent of computer workstations with powerful graphical processors has brought about a new class of graphics for consideration with respect to ISO 9001. This type of engineering graphics is the interactive three-dimensional or "solid" models. Although three-dimensional computer techniques have been available for decades, they were generally reserved for a few engineering specialists. Now, these systems are much more accessible and are more often found in the hands of designers as well as engineers.

Note that these models are not just a representation of static views of the part. These models can be manipulated. The part can be rotated, zoomed, panned, and sectioned. These models can also be used to create a virtual assembly of the entire product. This new graphical paradigm may require a greater input of effort to create, but it may prove beneficial in reducing ambiguities in the representations of product designs.

The use of computer models and virtual prototypes does present some difficulties with respect to ISO 9001 and Design Control. If a three-dimensional model is used to create rapid prototype parts (tangible parts created directly from the computer model by building up layers of material) and these parts are used in prototype testing, design verification can be provided and documented based on this testing (as in a Verification Study).

However, if the three-dimensional model is used with CAE or engineering software to predict the virtual prototype performance, and the computer model is considered the output from the Design Stage (with no Design Output Stage), following which the model goes directly into production, in this case only design validation would be occurring. In such a scenario, ISO 9001 may still apply, but all the design activities are computerized and the ISO 9001 Design Control section becomes just computer control. Furthermore, all computer systems and their software would have to be documented as valid and controlled. The engineering software would certainly have to be completely documented and verified to even attempt to show that Design Verification was occurring. In addition, the systems and software would have to be periodically revalidated against known product design benchmarks. This would be needed to show that a system, once it is perfected for designing via virtual prototype, is not inadvertently placed out of calibration.

Secrecy

Another important aspect of the design activity is secrecy. For the type of organization seeking ISO 9001 registration, the ability to design new products is likely to be a crucial aspect of its competitiveness. The value of the drawings, models, analytical methods, in-house engineering software, and materials used in products is paramount. These materials are often designated as trade secrets. This situation has an effect on a number of the sample forms shown in this chapter. Although protecting the intellectual property of an organization is not a direct concern for the quality of the designs produced by an organization, the organization should take full advantage of the procedures developed for ISO 9001 compliance to provide an appropriate level of secrecy as well.

All the graphical formats should have a confidentiality statement. At the very least, there should be a TRADE SECRET indicator on the pre-released drawing format (as shown in the upper left corner of Figure 5.20). In addition, a portion of the format may have a paragraph developed from the organization's legal department. This paragraph can indicate that the receiver of the prereleased drawing is bound to comply with the trade secret restriction (i.e., not copying or divulging information on the drawing). This legal information is essential to protect the investment that is made in the development of a new product or design.

A confidentiality statement or trade secret designation can also be added to the CAD sketch format (Figure 5.19) and the Engineering Sketch form (Figure 5.18). Although these documents may not be as likely to be passed to vendors for manufacture, they still contain valuable information that is owned by the organization investing in the product design. There also may be legal or security standards that apply to the organization, and designations such as "SECRET" or "CONFIDENTIAL" could be made part of the drawing format.

Note that some of the standardized documents presented earlier in this chapter may involve secrecy issues. The Verification Study for product design verification based on prototype testing (Figure 5.13) and the Verification Study for engineering standards or engineering software (Figure 5.5) should not be distributed outside the organization. In each case, critical design and engineering data are going to be documented and used. This information needs to be protected from inappropriate disclosure (accidental or deliberate). In order to provide some secrecy and the privilege of prosecuting misuse of the information, a confidentiality statement could be added on a page at the beginning of these documents. This statement might need to declare the classification of the document (such as containing trade secret or privileged information), the ownership of the document (stating that it is the property of the organization, not an individual), and a statement that the receiver is obligated to not copy or divulge the information (unless with proper written permission, perhaps).

Beyond a confidentiality statement in a Verification Study applied to engineering software, there are special secrecy considerations with regard to computer programs. In this case, the organization needs to show the user of the computer software that it is a proprietary work. When the programs are run or executed, a screen or statement should appear indicating that the program contains trade secret algorithms and that results from the program are trade secret information. Of course, this assumes that the program is developed in-house by the design and engineering organization.

The prevalence of personal computer systems and software also presents unique challenges. When centralized (mini or mainframe) computer systems were the norm, they could be configured to control the access of users to resources such as engineering software. Personal computers are often a different matter. Virtually by definition, one cannot furnish a computer program to a user without offering the means to copy the software. The magnetic media which often provides the software to the computer can be used to duplicate the software.

At the very least, engineering software developed by an organization for their design methods should clearly declare its classification. If it contains trade secret methods, then the program should clearly indicate to the user that the information in the program and the results from the program are the property of the organization. This does not prevent the software from being copied by personal computer users without permission, but it should give the organization the basis for handling the situation in the same manner as when trade secret drawings or documentation are passed to unauthorized parties.

Of great concern for the secrecy of engineering software is the source code. Although computer languages such as noncompiled or interpreted BASIC programming provides the user with the details of how the computer is programmed for a particular design method, other computer languages do not make these details as apparent. To help protect the methods contained in programs, most organizations should use languages that produce executable images from source code with a compiler. In this case, the user can run the software and produce results but cannot see the original computer programming method.

Regardless of the method of computer programming, a copyright and/or trade secret statement can be added to the source code. A sample statement for a program written in the C language is shown in Figure 5.21. Although this sort of procedure is not necessarily an issue for achieving ISO 9001 registration, ISO 9001 may be used as a reason

```
/******************************************/
/*                                        */
/*    (c) 1996     XYZ Corp.              */
/*    All rights reserved                 */
/*                                        */
/*  This source code, its object code, and */
/*  executable image are the property of XYZ */
/*  Corp. All unauthorized copies, compiles, */
/*  and executions are forbidden.         */
/*                                        */
/******************************************/
# include 'stuff.h'

main{

printf ("\n(c) 1996 XYZ\n  User to treat this program as Trade Secret\n" );


}
```

Figure 5.21   Sample computer source code secrecy segment.

for starting such a procedure if it is thought that the engineering software is not controlled or protected adequately. In the author's experience, such a course of action has been necessary in similar situations.

Final Remarks

Although ISO 9001 does not discuss how to design, there is going to be a varying degree of difficulty demonstrating the conformance of different design methods with ISO 9001. Some methods are going to be easily documented or verified, while other methods may introduce problems. This chapter has discussed the Design Stage of the ideal ISO 9001 product design process. It has presented a number of methods that should easily work in an ISO 9001 framework. The reader can likely effectively apply the information in the ideal forms and methods to their own organization's methods of product design.

References

1. ISO 9001 or ANSI/ASQC Q9001-1994 begins (paragraph 1, p. 1) with the statement "This American National Standard specifies quality system requirements for use where a supplier's capability to design and supply conforming product needs to be demonstrated." (See Bibliography.)
2. Schoonmaker, S. J., "Engineering Software Quality Management," *ASME Petroleum Division,* Vol. 43 (*Computer Applications and Design Abstractions 1992*), pp. 55–63.
3. Schoonmaker, S. J., "Engineering Software Quality Management and ISO 9000," *ASME CIE Division* (*Engineering Data Management Proceedings 1993*), pp. 195–204.
4. Schoonmaker, S. J., "Techniques in Engineering Software Quality Management," in Leondes, C. T. (ed.), *Control and Dynamic Systems,* Vol. 60-0, San Diego, Academic Press, 1994, pp. 289–328.
5. The reader should also refer to sources on software management in the Bibliography.

6

The Design Output Stage

Introduction

In the ideal ISO 9001 product development process, a stage is set apart for the creation of final and approved design information or documentation for actual release (to production and then the customer). This is called the Design Output Stage by the author, and this chapter is going to present some details about the activities in this stage.

Note that the distinction between the Design Stage and the Design Output Stage may not always be apparent, but there are advantages to making this distinction. The most basic intended advantage is to prevent extensive departmental procedures (at least those needed for the sake of ISO 9001) from distracting a design and engineering department from its most basic task—innovative product design. The Design Stage needs some controlled procedures, but the documentation used for product design with the Design Stage can be completely contained within the design and engineering department. Therefore the procedures can be controlled solely by that department.

The Design Output Stage documentation, however, has a different internal customer. This documentation must meet all the needs of the organization for product configuration. This internal customer satisfaction can include production control departments, production functions, technical publications, aftermarket functions, etc.

Recall that as the Design Output Stage is completed and all the final documentation is created and approved, the ideal ISO 9001 product development process included a design review activity. Although a design review activity has been presented within the Design Stage for design verification in the context of the basic functionality and technological feasibility of the design, this Design

Output Design Review activity needs to be performed to verify that the design output documentation is also satisfactory.

Other activities in the Design Output Stage are related to the Design Stage. In the two-dimensional paradigm, the designers may have created drawings (such as the prerelease format from the previous chapter), but they may still require detailing. In this Design Output Stage activity, the drawings are annotated and prepared for release. The release activity is also generally associated with preparing the documentation for manufacturing; however, there may be other internal documentation required for purchasing, accounting, material procurement, resource planning, etc. Another common Design Output Stage activity is the preparation of bills or bills of material (BOM).

Regardless of what kind of design output documentation is created, ISO 9001 makes special requirements for the design output documentation. ISO 9001 requires that this documentation be reviewed and approved prior to release. This approval activity may be ongoing during the design stage (with drawings approved as components are completed), but by the end of the Design Output Stage all the design output documentation is to be approved. ISO 9001 also requires that the design output documentation make some specific references. In particular, design output documentation is to reference acceptance criteria and identify those characteristics that are crucial to safe operation.

If a design input calls for a certain performance characteristic to be met (i.e., acceptance criteria), then the design output must show the design's value for that characteristic along with the value that is to be met by that specification. The design input documentation (or design specification) should have clearly stated these acceptance criteria, and the actual values to be compared to these acceptance criteria should be readily available.

The next ISO 9001 requirement for the design output is identifying and highlighting safety issues for the design. In some cases, the design input documentation may contain safety-related parameters. In this case, the design output would show this as an acceptance criteria. In other cases, a new product may have safety-related issues that are not considered at the time of the Design Input Stage. For instance, the new product may use processes or materials that have never been used before by the organization. If the safety-related issues have not been documented during the Design Stage, this must occur during the Design Output Stage.

Graphics

As presented in the Design Stage chapter, the design and engineering department is going to create and use a variety of graphical representations for the design (sketches, prereleased drawings, three-dimensional models, etc.). In the Design Output Stage, this department needs to create final and formal drawings. Although three-dimensional computer models may be approaching viability as the central and controlled database of product configuration, this situation presents some difficulties with respect to ISO 9001 (as discussed in Chapter 5). In order to focus on the impact of ISO 9001 in the Design Output Stage without the distraction of those special difficulties, the only graphical documentation discussed in this section is drawings.

First, it is assumed that the final or released drawings are based on the prerelease drawings used in the Design Stage. If a CAD system is used for the prerelease as well as the released drawings, there should be no need to recreate representations of the product (i.e., rework the part geometry in the drawings). There may, however, be a need to apply more detail to the drawings and add the final drawing format.

Dimensioning

One of the details which may be needed as the drawings are finished or detailed includes complete and proper dimensioning. This activity involves checking for consistency and applying standard appearance for the values shown as dimensions in the drawing (dimensions are shown in drawings as numerical values with arrows next to the part geometry as seen in Figure 6.1). The dimensions may need to have a standard approach applied (i.e., the values between separate lines with arrows or values next to unbroken lines). In addition to the values themselves, the unit system for the values (inch or millimeter for instance) needs to be addressed in the released drawing. The dimensions may also need to be detailed for tolerance values; in this case, the largest amount of variation permitted in the actual part is specified. For both the unit system and the tolerance information, default or standard information need to be added to the drawing (often as part of the released drawing format). Finally, with respect to dimensions, there needs to be a check for consistency. In some cases, the configuration of a part may be undefined if an insufficient amount of dimensioning is used. In other cases, the configuration of a part may be ambiguous if too many dimensions are shown. The full dimensioning activity is considered a Design Output Stage activity, and the standard procedures for this activity need to be documented and controlled as part of an ISO 9001 effort to show control over the design and development process.

Views

Related to the standardization of dimensions is the standardization of the views in the drawing. Views are related to the interpretation of the

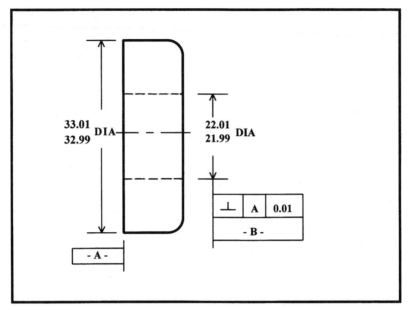

Figure 6.1   Sample dimensioning standardization.

part geometry and its dimensional information. These views have to be standardized and need to be clearly identified so that the receiver of the drawing knows how the part is being presented and dimensioned. This is a vital need for the organization and is essential for the achievement of quality in communicating the design to other parts of the organization and in the eventual satisfaction of the customer.

As shown in Figure 6.2, the views should be clearly identified in the release documentation or there must be a published and understood standard for use by the entire organization. Indeed, the use of drafting standards for the organization (not just the design and engineering department) is a definite advantage with respect to ISO 9001. The output from the Design Stage must be reviewed according to ISO 9001, and if there are controlled procedures or standards for this activity it should be easier to demonstrate the control over this activity.

Product data standards

Figure 6.3 shows a sample format for a drafting or Product Data Standard. There may be other forms of product configuration data such as bills of material, and the creation and control of this documentation may be handled by procedures under the general heading of Product Data Standards. These type of standards could be covered

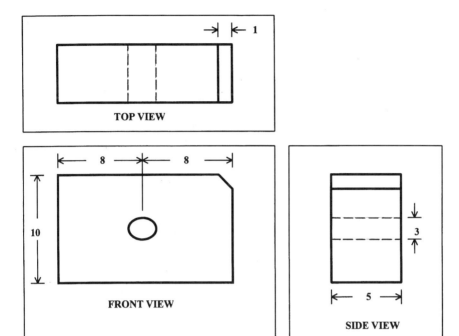

Figure 6.2   Sample view standardization.

in an Engineering Standard system (as presented in the Design stage for standardized design methods), but this information should be approved and/or reviewed outside of the design and engineering department to be certain that the needs of production or other departments are met satisfactorily. Therefore a separate Product Data Standard is envisioned.

The sample Product Data Standard shows a number and name identifier. The number and name would have to be cataloged and controlled. An index of these names and numbers should be available for search and retrieval of needed information. After this identification information, the standard should present a scope. The scope could indicate the types of product data (drawing versus bill, etc.) that the standard applies to; it could indicate what restrictions apply to the particular standard with respect to different markets (domestic versus export); it could also indicate what product or product lines are covered by the standard.

The next section of the sample Product Data Standard in Figure 6.3 is for References. In this section, standards or other documentation that is of interest or connection to the Product Data Standard are listed. With respect to the example of dimensioning on drawings, there is

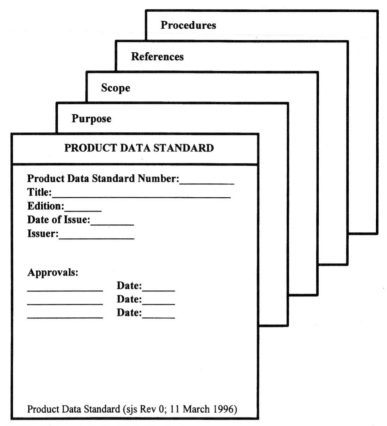

Figure 6.3 Sample Product Data Standard.

often a reference to ANSI standards (published by societies such as the ASME) or ISO dimensioning from the organization that publishes the ISO 9001 standard. Listing these types of standards in the References section can be helpful in understanding the basis of the organizations product data management. This, in turn, provides another facet of demonstrating that the information from the Design Output Stage is being properly controlled and approved.

The next section of the sample Product Data Standard covers the Procedure. For the example of the standardization of views on released drawings, this section could specify the placement of the views on the drawing (where the Top, Front, Side, etc. views are to be placed on the drawings) and the identification of those views (how the views are to be noted in the drawing or whether an isometric or composite view can be shown and how it would be identified as such). Finally, the sample Product Data Standard contains a section on the first page for approvals. The approvals for these standards should include all the departments affected by the product data information such as the standard views. As is expected in the Design Output Stage, this standard is not just for the use of the design and engineering department.

Note that the issues such as dimensional tolerance information and the standardization of the view representations are vital for passing the correct product design information to the production environment. Some other issues related to production include installation instructions (which often include graphical information and may be contained within released drawings), manufacturing processing instructions (such as surface finish, tooling requirements, casting and fabrication methods specifications, etc.), and material specifications (either an internal material identification or industry standard identification). All these issues need to be controlled, and a Product Data Standard system of some kind can be used to provide this control. For ISO 9001, all these issues must be satisfactorily addressed in the Design Output Stage. This should be satisfactorily addressed by building and conforming to a set of Product Data Standards. These standards (as with the engineering standards) would be a lower level of documentation referenced by the quality manual.

In addition to production related aspects of the released drawings, there may be information related to procurement. Some parts specified in released drawings are not going to be produced by the design and engineering organization. Instead, these parts may be produced by a vendor and the part information then needs to be passed to the vendor in a controlled manner. In this case as well, there could be a Product Data Standard which covers the procedure for vendor documentation. And again, the organization is going to need to demonstrate conformance to that procedure for the ISO 9001 effort.

Drawing format

The final aspect of graphics for the drawings created in the Design Output Stage is the drawing format. As discussed in Chapter 5, the drawing format is the standard textual information usually found at the border or edges of the drawing. Figure 6.4 presents a typical format for the Design Output Stage. Note that this format contains what is referred to as a *title block* which is the area of textual information at the bottom of the drawing which includes Title, Part Number, etc.. This drawing format also introduces issues for drawing review and approval, revision levels, secrecy, and part numbering. Each of these issues are addressed with respect to the drawing format.

First, the drawing format needs to be standardized for the organization. The drawing format may also be developed in accordance with

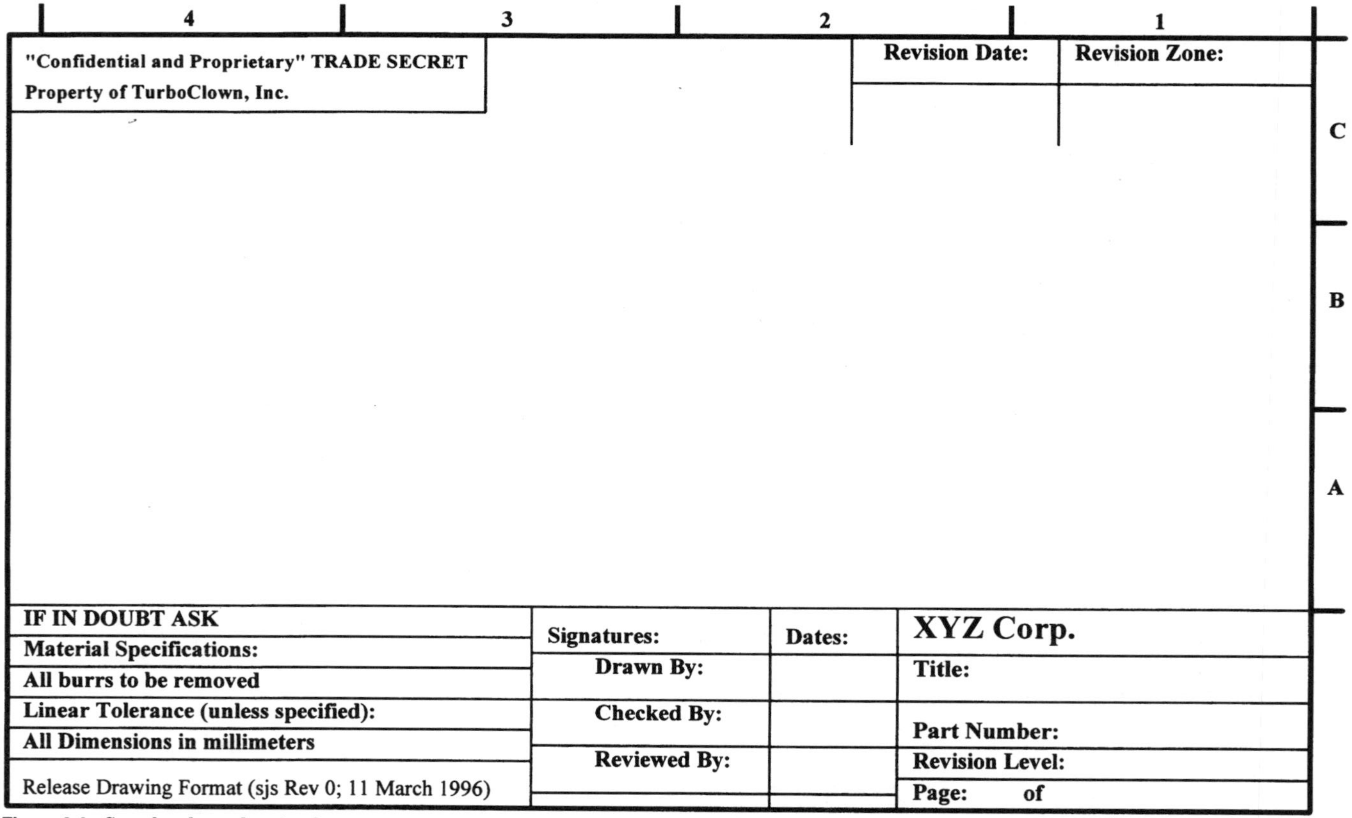

"Confidential and Proprietary" TRADE SECRET
Property of TurboClown, Inc.

	Revision Date:	Revision Zone:

4 · 3 · 2 · 1

C
B
A

Signatures:	Dates:	**XYZ Corp.**
Drawn By:		Title:
Checked By:		Part Number:
Reviewed By:		Revision Level:
		Page: of

IF IN DOUBT ASK
Material Specifications:
All burrs to be removed
Linear Tolerance (unless specified):
All Dimensions in millimeters

Release Drawing Format (sjs Rev 0; 11 March 1996)

Figure 6.4   Sample release drawing format.

industry standards (such as ANSI or ISO). Thus, in order to maintain standardization, a Product Data Standard should be created that specifies the drawing format and its information.

Second, the title block shown in Figure 6.4 includes much of the information of relevance for ISO 9001. Obviously the title of the drawing is shown in the title block. This should be the basic description of the part shown in the drawing. In many cases, however, the number identification is more central to the identity of the drawing and the part contained within it. In this case, the number shown on the drawing is called Part Number. This number should be used to define the part in the organization's database of parts. In order to preserve the integrity of that database, the part numbers obviously need to be carefully controlled. There needs to be a master list of part numbers, and the disposition of each number needs to be known at all times. With respect to the activities of the Design Output Stage and achieving success with ISO 9001, the assignment or checking of these numbers is going to be essential.

If a set of preliminary or special prerelease drawing numbers are used during the Design Stage (in order to permit greater flexibility in creating, changing, and dropping parts and assemblies), then the Design Output Stage needs to review and organize the drawings from the Design Stage using final or released part numbers.

The next item of the drawing format in Figure 6.4 to be discussed is shown as Revision Level. This refers to the version of the drawing (and perhaps the version of the part shown in the drawing). Preferably, the Revision Level indicates the number of times the drawing has been released from the design and engineering department. Note that there may have been changes to a prereleased drawing during the Design Stage, but once the drawing is in the Design Output Stage those changes could be ignored (since those drawings are assumed not to have been released in any case). Regardless of what system is used by the organization to control the revision level of the drawing, a Product Data Standard should be created to specify the system and this Product Data Standard needs to cover the use of the Revision Level as an item in the drawing format.

When the Design Output Stage is complete, the Revision Level of the drawings must be known. For new drawings created for new parts, this Revision Level can be set to a consistent value (A, 0, or 1 as desired) as dictated by a Product Data Standard. Of course, a new design may share drawings or parts with existing, released designs. In this case, the drawings could be already sufficient or the drawing could be revised to meet the needs of the existing design and the new design. In any case, the Design Output Stage would have to include an activity for the revision of the released drawings that need to be changed to reflect the new design. The revision level of modified drawings is simply incremented (from A to B, or 1 to 2, etc.). As usual, this activity should be in accordance with a Product Data Standard.

With respect to the drawing format in Figure 6.4, an area of the format separate from the title block may be included in the format to detail the revision activity. This is shown in the upper right corner of the drawing. For each Revision Level over the life of the drawing, there should be shown a description of the change made to the part or the drawing. In many cases, this description includes a reference to a zone of the drawing and the zones and boundaries may be shown in the drawing format border or edge. Once again, this activity would be covered by a Product Data Standard for revision level.

The next item in the drawing format of Figure 6.4 to be discussed is Page (at the lower right corner of the format). This section of the title block should simply indicate the number of the page or sheet attached to that title block. When more than one page is needed to properly show the configuration of the part, more pages may be used. In this case, the page number and the total number of pages associated with the part number needs to be shown on the drawing. This is needed to let the receiver of the drawing know whether all of the part information is being accessed. The setting of the page numbers into this drawing format may be an activity in the Design Stage, but the Design Output Stage needs to check that this information is correct with respect to release to production and other customers.

The next items in the drawing format are for Drawn By and Reviewed By with dates for each. These items are found in the center bottom section of the format in Figure 6.4. These items refer to the name of the person that creates the drawing and the name of the person that reviewed the drawing.

The review (or approval) of the drawing is an ISO 9001 requirement, since ISO 9001 specifies that design output documents be reviewed. Although, the design review meeting at the end of the Design Output Stage can provide the review requirement for drawings, there may be a large number to be reviewed (dozens, hundreds, or thousands of drawings). A design review meeting may not be an appropriate format for that activity. Instead, the drawings can be approved by authorized personnel prior to the design review meeting.

The person providing the approval for the drawing needs to consider the basic geometric configuration of the part, the materials for the part, as well as the Part Number, Revision Level, and other product database information. The compliance of the drawing appearance to standards for views and dimensions, etc. should be provided by the person indicated by the Checked By item.

Some organizations may want to connect the record of design verification to the title block. In this case, the design specification acceptance criteria and calculated and/or measured values for the part or assembly can be included in the drawing format. The designer making the design verification calculations (by alternative methods or by prototype testing) could then add his signature to the drawing under a Verified By item. Of course, if this method is chosen then the procedure needs to be clearly stated in the Product Data Standard.

Considering the importance of the drawings from the Design Output Stage for producing correct parts, the personnel providing the review of the drawings should have actual signatures on file (assuming the drawing itself is not physically signed). In many cases, the drawing format is contained in a CAD system, and the reviewer's name for the drawing is going to be typed into the area of the title block for Reviewed By (as opposed to physically signed or initialled by the reviewer). In order to prevent any questions about the validity of the reviewer with the electronic format for the drawing, the names and signatures of the persons that review the drawings should be kept in a file system specified by a Product Data Standard. These files can be referenced as needed during an ISO 9001 audit situation.

Another issue related to drawings and the drawing format is secrecy. Secrecy is an issue for all documentation that contains proprietary technical information. This is certainly the case with drawings which are going to contain trade secret information. To clearly identify drawings from the Design Output Stage as confidential and as containing trade secret material, a section of the format should be devoted to a confidentiality statement. It should state that the drawing is the property of the design and engineering organization, that the receiver of the drawing may not reproduce the documentation or divulge any information from it (without permission), that the receiver must return it to the originating organization, and other statements as needed to protect the organization's drawings from improper disclosure.

Technical publications

In most cases, the production and control of release drawings is going to be a major focus of the Design Output Stage. In addition to these drawings, other types of graphics may be prepared in the Design Output Stage. This may include drawings of a general nature not associated with the part numbering system (such as general arrangement drawings). This might also include drawings for technical publications (such as training manuals, maintenance manuals, parts lists, etc.).

This auxiliary graphical information may need to be controlled in the same manner as the released drawings. Although this graphical information may not be released to a production department for use in the procurement and production of the product, and it may not be seen as a direct contributor to the product's performance, it is still important. In terms of the ideal ISO 9001 process, if the timely creation of this graphical information is a customer requirement, deliverable, or acceptance criterion, then this requirement must be met just as much as a performance requirement for the product.

Therefore, to be certain that the ISO 9001 requirements are being met, the associated graphical information should be controlled in the same manner as the release drawings. General arrangement drawings and technical manuals should be given numbers or other identifiers, there should be revision level applied to the documents, and reproduction and distribution should be serialized and controlled with master lists.

Finally, in case operators' manuals or other formats are widely distributed to customers and operators of designs, the associated graphical information may be protected by a copyright instead of a trade secret designation.

Document Control

As a basic consideration, all the documentation from the Design Output Stage (not just drawings) should be considered controlled information. As such, certain attributes must be associated with the documentation. This documentation should be based on a master source of information such as a master blueprint or document, or preferably a computerized database of text and graphical information. Obviously, this centralized computer information would need to be based on computer equipment with a proper environment, controls, and adequate safekeeping or backup procedures.

Once the master source of information is created, the information needs to be provided to appropriate users in a controlled fashion. In a paper-driven system, copies need to be created and the copies numbered. Based on a master list of recipients of the documentation, the copies would be distributed and maintained. In this system, the source of the controlled copies are assumed to be automatically updated so that whenever someone references that document, the current and correct information is obtained. Of course, this can be difficult to maintain in a paper-driven system since the recipients must provide the maintenance manually.

In a computerized documentation system, the information can be maintained automatically. When documentation (such as drawings and documents from the controlled Design Output Stage) is created, then the computerized database of text and/or graphics contains the current information. The recipients of the information, then simply become viewers of this information in the on-line system; there is no need for manual maintenance.

In this system, copies of the on-line information should be discouraged. The on-line reference should always be sought. There may be cases where paper copies are needed, but these should be clearly marked with an UNCONTROLLED stamp, and they should be destroyed after their use. Of course, the problem of copies made from the master or controlled documentation may be even more acute in the paper-driven system.

As covered in the ISO 9001 paragraph 4.5 on Document and Data Control, any system used for control of documentation such as Design Output Stage documentation must provide adequate coverage. The organization must determine what persons or groups require the controlled documentation and see that these areas are covered for access to the information. The system must also prevent the use of obsolete information. Finally, if obsolete or superseded design output information must be retained for legal reasons, this information must be segregated from the current information to ensure that they are not mistakenly used for production purposes.

Record retention

Another aspect of the documentation control is record retention. This is a policy that the organization (or at least the design and engineering department) must create and maintain which indicates how long obsolete information must be retained. This period is likely to be related to the legal and liability issues for the organization. It may also be related to the organization's amount of activity in the aftermarket or parts sales and product support business. In many cases, as a design is released to production and eventually to the customer, the design is refined, expanded, and enhanced. As customers of early versions of a product design may need aftersales support, the design output documentation may need to be available. However, as with the retention of this information for legal reasons, this documentation must be successfully segregated to prevent accidental use in the production of the latest or current model of a product.

Whatever amount of time is deemed desirable or necessary for record retention, it should be documented and approved. The Product Data Standard system presented earlier for drawing and part numbering standards could easily be used for this purpose. Figure 6.5 presents a sample record retention Product Data Standard. As would be expected, the current or active information would be retained indefinitely. However, as information is superseded or made obsolete, it

---

PRODUCT DATA STANDARD

Product Data Standard Number: __PDS1001__   Edition: __1st__

Title: __Record Retention__   Date of Issue: __3/12/96__
                                                   Issuer: __S. Schoon.__

1.0 Purpose - Set the policy for retention of quality system documentation

2.0 Scope - This standard applies to the design and engineering department. This standard
             takes precedence in any department without a specific policy for product data.

3.0 References - ANSI/ASQC Q9001-1994 4.5 Document and Data Control
                  XYZ Corp. Standard QC1001 Document Control

4.0 Procedures-

   4.1 Following classes of information to be retained as follows:

            Released but superseded drawings - 5 years after superseded or as dictated by
                                               legal requirements
            Pre-release drawings - 5 years after product release
            Project data files - 20 years after product release
            Superseded Engineering Standards - 10 years after superseded
            Superseded Product Data Standards - 5 years after superseded
            Superseded Engineering Software - 10 years after superseded

   4.2 This information is to be displayed in plain view in all areas containing these
        classes of information.

   4.3 Prior to disposal, a notice of the documents to be destroyed is to be sent to
        the Director of Engineering.

   Approvals:
   _____ Date:_____
   _____ Date:_____
   _____ Date:_____

   Product Data Standard (sjs Rev 0; 11 March 1996)

Figure 6.5 Sample record retention.

would then be retained for an amount of time as specified in the standard.

Although some of these issues may be beyond the scope of the Design Control paragraph of ISO 9001 (and into other paragraphs of the standard), they are still often of concern to designers and engineers. This is the case since designers and engineers are often the creators of the documentation that needs the greatest amount of control. Although the control, distribution, and maintenance of this documentation may be provided by an engineering services department within an organization, designers and engineers should still be aware of the needs of the system.

Bills of Materials

Another common activity in a Design Output Stage is BOMs or scope of supplies. Usually activities related to this issue involve organizing the product database with respect to in-house production versus vendor production, or organizing with respect to use of commodity or general-purpose versus special part numbers, or organizing with respect to assemblies of parts versus single or piece parts. In addition, there is usually a hierarchy of BOMs related to parts, subassemblies, assemblies, master assemblies, vendor assemblies, etc. This hierarchy then determines the how parts are arranged to create the final product configuration.

The issue of bill of materials or BOM may be considered an ancillary task for some designers and engineers (as in the case of document control issues presented earlier). It may not be perceived as being as important as the activity of providing innovative concepts and products. In some cases, the BOM activities are handled by an engineering service department, or they may be handled by an area of the production control department. As such, they may not be a direct concern of designers and engineers. However, as with document control, designers and engineers are often the creators or sources of the information which must be incorporated into the BOM. Designers and engineers therefore should be involved with the standardization and development of the policies related to the BOMs. Although creation of a well-conceived and controlled BOM may not be a customer requirement indicated as an acceptance criteria in the design specification, it clearly needs to have the input of designers and engineers to ensure that the product as conceived and tested becomes the product that the customer receives. Since meeting customer requirements is central to ISO 9001, the creation and proper control of BOMs should be considered a vital issue for having the design and engineering department conform to ISO 9001 (regardless of whether it would be considered part of the Design Control, Document and Data Control, or Process Control paragraphs).

As with the database of drawings in a CAD system, a computerized database may be used to great effect for BOM (perhaps in conjunction with a production control software system or even a CAD software system). In this case, parts are identified by part number and organized into groups such as widely used and rarely changed items such as hardware. These parts could then be considered commodities that can be called out from a variety of designs, products, and product lines. In the computerized database, the part number can then be used to locate the release drawing (for manufacturing or procuring the part) or location in inventory (for accessing existing parts). Other parts can be segregated as peculiar or special to a specific design, product, or product line. Again, in the database, the part number can then be used to locate drawings and/or location in inventory.

In addition to considering the organization of part numbers, the BOM system needs to consider the creation of assemblies and quantities of parts. Although designers and engineers may spend a majority of effort on the geometric configuration of parts within drawings, the final product is likely to be an assembly of many parts and some of those parts would be repeated in the product at different locations. Thus, the part drawings are insufficient to completely describe the design. The specification of the quantities of parts and their association with other parts can be addressed and maintained with the BOM database. Each part or drawing in the product database needs to indicate what parts and quantities of those parts are needed to be added to the part shown. Each part, in turn, may be a child of another part, and so on, until there is a top level that starts the hierarchy. Figure 6.6 shows a sample report for this type of information that might be generated from bills of materials.

Since the BOM information is considered part of the Design Output Stage in the ideal ISO 9001 development process, and design output documents must be reviewed before release, there must be review of this information. One possible avenue for approval of the bill is to produce a report or other paper document with the all the part numbers, quantities, and hierarchy shown. This report or document could then be signed by an appropriate manager or expert to demonstrate that the information has been reviewed.

Unfortunately, the BOM may change during the life cycle of the product as commodity type of parts may be changed independently of the peculiar parts specifically associated with one product design. Thus the approval is only valid at the time it is reviewed. Probably a better approach for review of BOM information is to include its review as part of the design review activity at the end of the Design Output Stage. In this case, the BOM, part numbers, and hierarchies can be reviewed at that point in time. From that point, changes to a BOM would have to be requested and executed in a controlled fashion (as in a Engineering Change Request format and an Engineering Works Order for revising bills of materials). Subsequently, before revisions to a BOM would be released into an active computerized

		BILL OF MATERIALS (TOP LEVEL) (reference Product Data Standard PDS2001)		

Prepared by:_____   Date:_____   Data Software/version:_____

Level	Qty	Parent P/N	Entity	Child P/N ('s)
0	1	Product Code X1000	Assembly, Top	A0000,B0000,B0001
0-1	1	A00000	Assembly, Vendor, Std	*
0-2	1	B00000	Assembly, Particular	B00100,B00111,...
0-3	10	B00001	Part	B00200
0-4	1	D00000	Manuals	D00100
. . .				
1	1	B00100	Sub-Assembly,	B00201
1-1	2	B00100	Part, Vendor, Non-Std	B00112
. . .				
1-20	4	D00100	Manual, Instruction	*
1-21	2	D00100	Manual, Maintenance	*

Bill of Materials Format (sjs Rev 0; 12 March 1996)

Figure 6.6 Sample bill of material.

database, the output from the Engineering Work Order would have to be reviewed and signed by an appropriate manager or expert.

Although a paper-based system may still be an alternative for a BOM, most organizations covered by ISO 9001 are going to use a computerized system. There should be no problem with using the computerized system in an audit situation. The Document and Data Control portion of the ISO 9001 states that documents and data can be in the form of any type of media (such as electronic media[1]) found in an online database.

Of course, a computerized system needs to be given a secure environment, controlled by documented procedures, and backup systems must be in place. These are the issues that may present problems in an audit situation. Regardless of the level of reluctance for documenting and controlling the fast-paced arena of information technologies, basic integrity of the system that contains vital product information must be maintained at all times and this may need to be demonstrated in an audit situation.

Project Files

The next aspect of the Design Output Stage to be discussed is the preparation of a Development Project File. This file is a collection of documentation (electronic as well as printed media) that should be retained and maintained by the design and engineering department at the end of a development project. Some of this information may be retained for legal reasons; other information may be retained for use as reference data for future development projects. In either case, the information in the Development Project File would not be released information (such as drawings, BOMs, etc.). Although this information is gathered and organized in the Design Output Stage, the information is reference information that is not expected to be passed to the production or manufacturing engineering department.

Although a Development Project File may not be required by the ISO 9001, it can be helpful in demonstrating compliance with ISO 9001 since all the important documentation is compiled and controlled together. Throughout this discussion of the ideal ISO 9001 development process, a number of documents are specific requirements for ISO 9001. These documents are the Design Specification (Figures 4.1 and 4.2), the Verification Study (Figure 5.5), Engineering Test Report (Figure 5.12), Prototype Verification Study (Figure 5.13), the Design Review Notice (Figure 5.16), the Design Stage Design Review (Figure 5.17), and the Design Output Stage Design Review (Figure 6.7). If these documents are controlled from a number areas or groups, then there may be more difficulty in demonstrating compliance.

Thus, these documents can be brought together in the Development Project File to assist with an ISO 9001 registration effort. All these documents are going to be generated by each of the development projects and each of these projects should have an identification such as the new product code. Therefore, the project data can be gathered and filed based on that development project identifier. This information should then be traceable based on that development project identifier. Recall that the identifier is also tracked by the Development Project Master Log (Figure 4.4). There should be a Development Project File for each of the projects shown in this log file.

```
┌─────────────────────────────────────────────────────────────┐
│  ┌───────────────────────────────────────────────────────┐  │
│  │        DESIGN OUTPUT STAGE ("FINAL") DESIGN REVIEW      │  │
│  └───────────────────────────────────────────────────────┘  │
│                                                               │
│     New Product Code:_____      Date of Review:_____     │
│     Project Engineer:_____ (e-mail address:_____)          │
│                                                               │
└─────────────────────────────────────────────────────────────┘
```

	Check if complete:
Design Verification:	☐
Safety Review:	☐
Regulatory Review:	☐
Drawings (format & approvals):	☐
Bill of Materials Review:	☐
Project Data File:	☐
Manufacturability Review:	☐
Materials Review:	☐
Performance Review:	☐
Environmental Review:	☐

Approvals:

_____ Date: _____
_____ Date: _____
_____ Date: _____

Design Output Stage Design Review (sjs Rev 0; 12 March 1996)

Figure 6.7 Sample Design Output Stage (Final) Design Review.

In addition to the documentation needed for ISO 9001, various reference information (and information needed for legal reasons) may be included in the Development Project File. Some examples of this type of information include the standard Calculation Form (Figure 5.2), Engineering Data Recording Sheets (Figure 5.8), the Engineering Test Log (Figure 5.9), Engineering Test Request (Figure 5.10), Inspection Problem Incident Sheet (Figure 5.14), Problem Incident Log (Figure 5.15), and Prerelease drawings (Figure 5.20).

Some of this information could be maintained in other systems besides the Development Project File. For instance, the engineering test information (data recordings, test logs, and test requests) could be maintained in files associated with the testing laboratory or facility. For handling such a case, the sample Engineering Test Request

Log (presented in Figure 5.11) cross-references the new product identifier (such as the New Product Code) with the Test Number.
Therefore, the project data involved with testing could be kept in a
separate file system since the information can be traced when needed
(based on the Test Number in the testing file system).

Another area for consideration for inclusion in the Development
Project File is the designer's or engineer's Design File shown in
Figure 5.1. In some cases, this information may be deemed too preliminary to include in the Development Project File. The sketches
from this stage are probably documented in prerelease drawings anyway. There may be no need to repeat this information or its unsuccessful design configurations. Also, information from the engineer's
Design File may be too esoteric (such as literature search information); this material may not be worthy of inclusion in the
Development Project File.

Of course, the unsuccessful configuration and literature search
information may be quite valuable in the future to other designers
and engineers in the design and engineering department. If the individual designers and engineers wish to collectively organize and
maintain the Design Files generated by their department, then obviously this information could be organized and maintained by moving
the Design Files into the Development Project File when projects are
completed.

Design Review

As part of the ISO 9001 ideal product development process, a design
review activity is placed at the end of the Design Output Stage. As all
the released drawings, bill of materials, project file, etc. are completed, a review meeting with representatives from all functions related
to the design (including manufacturing since this department is certainly related to the design as the basic customer for the release documentation) would be conducted. Depending on the development project, separate meetings could address different areas of the Design
Output Stage, or separate meetings could address types of output
(drawings, BOM, etc.). This Design Output Stage Design Review can
be considered a Final Design Review as mentioned by one of the
guideline documents associated with ISO 9001.[2]

The procedure for initiating, conducting, and documenting this final
design review activity could be the same as the design review used for
verification of design in the Design Stage (the Design Stage Design
Review). As discussed in the ideal ISO 9001 process, the design review
in the Design Output Stage needs to compare the release or final docu-

mentation with the design specifications (or compare the design output with design input and check that acceptance criteria, safety, and regulatory requirements are met by the design output).

In light of the demands of the design output to be documented and approved, and specifically in light of the reference design input acceptance criteria, an organization may decide to create and control a special final design review document. If the organization feels that all the Design Output Stage documentation requirements are met by existing drawings and reports with their approval signatures, this material would not be needed.

A sample Design Output Stage Design Review form is shown in Figure 6.7. This sample document shows a checklist for the meeting results. The first item to check is Design Verification. If a Design Stage Design Review has been conducted, and the minutes or standard form (see Figure 5.17) from that meeting is correct and in the Development Project File, then this item can be checked as completed.

The document in Figure 6.7 can also contain a checklist for applicable regulatory, safety, and environmental requirements. As in the Design Stage Design Review, these issues should be discussed in a proactive manner and scenarios reviewed for the proper use of the product.

Unlike the Design Stage Design Review, this Design Output Stage Design Review needs to review the documentation from the Design Output Stage. The drawings should be in conformance with the Product Data Standard and should have been approved already; the BOM should be shown as reviewed and approved, etc. The Development Project File should also be reviewed for completeness. Finally, the document in Figure 6.7 can contain a sign-off section for clearly demonstrating approval of the final design output documentation.

Conclusion

This chapter has presented some discussion and detailed information about the Design Output Stage which is perceived to exist in the ideal ISO 9001 design and development process. The boundary between this stage and the Design Stage may not always be apparent. However, the Design Stage activity is considered to be finished when a design review meeting in the Design Stage is conducted and approved. At that point, the product is considered configured and this Design Output Stage is commenced to formally prepare the documentation for the design for release or passing on to the internal customers.

Some of the documentation that is completed in the Design Output Stage includes formal drawings, BOMs, and a project completion or archive file. All this material is considered crucial to success with an

ISO 9001 registration effort, and is assumed to be controlled by procedures that meet paragraph 4.5, Document and Data Control, of the ISO 9001 standard.

As with the previous chapters, the formats and forms and procedures presented in this chapter are primarily meant as a guide for helping the reader succeed in an ISO 9001 registration effort with his or her own organization.

References

1. ASQC, *ISO 9001 (ANSI/ASQC Q9001-1994) Quality Systems—Model for Quality Assurance in Design, Development, Production, Installation, and Servicing,* ASQC, Milwaukee, Wisconsin, 1994, paragraph 4.5.1 or note 15, p. 4.
2. ASQC, *ISO 9004 (ANSI/ASQC Q9004-1994) Quality Management and Quality System Elements—Guidelines,* ASQC, Milwaukee, Wisconsin, 1994, paragraph 8.6, p. 10.

The Design Validation Stage

Design Validation Stage

The last stage of the ideal ISO 9001 design and development process is the Design Validation Stage. As discussed earlier, the Design Validation Stage follows the Design Output Stage. It is an assessment stage where a decision needs to be made concerning the new product created from the new design. In particular, the organization must decide if the product design does in fact meet the customer's requirements (as opposed to purely just meeting technical design specifications).

Design validation is not the same as design verification (at least in the author's interpretation of the ISO 9001 standard). It is assumed that design verification is generally a technical activity for engineers and designers within the Design Stage already discussed. It addresses the issue of whether the output from the Design Stage meets the design specification. In contrast, design validation is generally an activity for management and marketing personnel in the context of the customer's original request for a new product (or the perceived market demand).

Of course, design validation activities may or may not involve technical judgment. A complex product design may have sophisticated technical customer requirements or performance envelopes. Without a technical background, it may be difficult to appreciate the performance and specification information. Although the Design Validation Stage may not involve technical judgment in some cases, it should still be preferable to use an interdisciplinary approach that includes technical expertise. In this approach, designers and engineers as well as marketing and project management personnel would agree on the design validation. As a general rule, design validation is not going to answer the question of whether the design works. Instead, the purpose of design validation is to answer the question of whether the design is going to satisfy the customer.

Quality Record

Considering the importance of this decision to the ISO 9001 ideal process, there must be a quality record or document created for demonstrating that the Design Validation Stage has been successfully completed. Furthermore, this document should be reviewed and approved by an appropriate level of management.

As usual, a standard form can be created by the organization for design validation. This form should indicate the project or design in question and the contract or other documentation that shows the customer's requirements. This document could also make reference to the design input or specification document and the design output documentation. The form can then state that the design has been assessed from the perspective of the customer, and that the design is ready for release to the customer. The design validation document may also be used to indicate a rejection of the design. Most likely this would be followed by a return to the Design Input Stage (to alter the specifications) or a return to the Design Stage (to alter the design).

Figure 7.1 shows a sample Design Validation form. The format of the document includes the project identification such as the New Product Code. The project engineer as well as the representative from the marketing department are indicated on the form. These persons would be considered as responsible for conducting or at least coordinating the design validation activity.

The sample Design Validation form next shows a Contract Number. This sample assumes that a design is created on the behalf of a specific customer under the rules of a contract for services. If a product is developed to meet a perceived demand in a general market (as opposed to a specific product under a contract), then a document from the marketing document (such as a New Product Proposal in Figure 4.3) specifying the perceived demand could be referenced.

In order to create a single quality record for design validation, the form shows a format for listing the customer requirements and then adding a check or tick to indicate that those performing the design validation function have determined that the customer requirement is met. If an organization does not wish to relist the requirements that are already listed in the contract or marketing proposal, then a paragraph can be placed on the form that states those performing the design validation have decided that the design meets the customer requirements as stated in the contract or proposal.

```
┌─────────────────────────────────────────────────────────────┐
│                    DESIGN VALIDATION                         │
├─────────────────────────────────────────────────────────────┤
│  Proposal Number:_____ / Contract:_____           │
│                                                              │
│  New Product Code:_____      Date:_____                │
│                                                              │
│  Project Engineer:_____      (e-mail address:_____)   │
│  Project Manager (Marketing):_____  (e-mail address:_____) │
├─────────────────────────────────────────────────────────────┤
│                                                              │
│  Customer Requirements:                    Check if met:    │
│      1:_____   ┌──┐             │
│      2:_____   ├──┤             │
│      3:_____   └──┘             │
│                                                              │
├─────────────────────────────────────────────────────────────┤
│  Check one of the following:                                 │
│                                                              │
│   ┌──┐  Design is validated for release   ┌──┐ Design is not validated due to │
│   └──┘                                     └──┘ requirements not met above     │
│                                                 or due to the following:       │
│                                                 _____        │
│                                                 _____        │
│                                                 _____        │
│                                                                                │
│  Approvals:                                                  │
│   _____  Date: _____                               │
│   _____  Date: _____                               │
│   _____  Date: _____                               │
│                                                              │
│                                                              │
│  Design Validation (sjs Rev 0; 12 March 1996)                │
└─────────────────────────────────────────────────────────────┘
```

Figure 7.1 Sample Design Validation Form.

Since the validation of the design may fail, the sample form in Figure 7.1 includes a means for rejection. In this case, those responsible for design validation check the appropriate box and then explain why the validation failed. Regardless of whether the design validation is complete or rejected, those responsible for design validation should sign the form to indicate that they agree with the results of the process.

In a large development project, such a simple form may be inadequate. A separate design validation process may be needed for each component alone or subassembly of a product. The process of validation may involve different personnel for the different areas of the design. Each of these validations would then need to be recorded and controlled for future use in an audit situation.

Finally, the quality record or records from the Design Validation stage may be added to the Development Project File. This may be useful for accessing all the ISO 9001-related information in a central location in an audit situation.

Testing

Even when the ideal ISO 9001 development process presented is used, the amount of time and effort spent in the Design Validation Stage can vary widely. In some organizations and types of products, a simple form (such as the sample in Figure 7.1) can be created and approved based on the design output documentation and the customer's requirements or contract. In other cases, the organization may perform a series of tests and demonstrations with a new product based on the new design.

In yet other cases, it may be impossible to perform realistic tests before shipment of the product to the customer. In the author's experience, this has been the case for the gas compression industry. New compressors are engineered for a set of customer conditions (inlet and discharge pressure and flow rate) for virtually any known gases (such as ethylene at pressures of 60,000 pounds per square inch). In many cases, the gases to be compressed are simply not available at the production or testing facilities. The design and engineering organization could then consider using an alternative gas for testing in the Design Validation Stage, or the new product could be sent to the customer site for field testing. Of course, testing the new design at the customer's site (field testing) violates the ideal ISO 9001 philosophy that is being presented, where the checking of the design is to be done within the confines of the design and engineering organization.

It should be noted that the purpose of the Design Validation Stage is to deal with a version of the new product design after a process of production. At the end of the Design Output Stage, final design documentation is released to the production department. In a so-called preproduction or production test process, a pre-full-scale production model is then assumed to be created using the anticipated manufacturing processes (as opposed to a prototype or new concept model). It is this production test model which is assumed to be available at the Design Validation Stage.

If testing is performed at the Design Validation Stage using the production test model, ISO 9001 provides some considerations for this process. These considerations include validating under defined operating conditions (such as the end-user environment), validating for multiple uses, and producing proper documentation for the validation process.

It is assumed that a production test model is used for testing in the end-user environment (if possible). Although prototype model testing may be done within the Design Stage (within the perspective of feasibility and performance), design validation testing is done with a production model from the perspective of the customer. The customer perspective testing can then be tailored to demonstrate that the design meets customer requirements. As usual, the end result must be a document that clearly indicates whether this testing was successful or not. Also, the document should be reviewed and approved by an appropriate level of management, that is, by personnel familiar with the design who have the authority to take corrective action if needed.

Another ISO 9001 consideration for testing in the Design Validation Stage covers multiple uses for the product. Presumably some new products or new concepts could be based in multiple applications by the operator or customer. If this is the case (particularly if such use is covered by a contractual agreement), the production test model would be tested in those different regimes. For each regime or type of use, the usual design validation documentation would be prepared, reviewed, and approved.

Variations in the forms and procedures presented in the Design Stage for engineering and prototype testing should be applicable to the design validation testing as well. Most likely, the validation testing is going to be performed by the same group of personnel that has performed those other types of tests anyway. Therefore, it may be best to use the same organization of procedures and reports for the results of design validation testing. There could be a Validation Test Request based on the Engineering Test Request (Figure 5.10), Validation Test Procedures based on the format of the Engineering Test Procedure (Figure 5.7), and a Validation Test Report based on the Engineering Test Report (Figure 5.12).

In addition to the engineering test type of documents, the prototype testing formats may be applicable as well, particularly if the production test model is inspected (from the customer perspective in this case) for the Design Validation Stage. The formats that may be useful in this regard include the Inspection Problem Incident Sheet (Figure 5.14), the Problem Incident Log (Figure 5.15), and the Prototype Verification Study (Figure 5.13). In the case of the verification study, a Validation Study or Production Test Study would need to be created based on the Prototype Verification Study. Instead of presenting alternative calculations for the new design, the Production Test Study would simply reference the production test product and the results of the inspection. These validation inspection documents can then be kept in the Development Project File for easy reference in the ISO

9001 audit situation, or they can be kept in the testing facility file system (as discussed previously in Chapter 5).

Benchmarks

In addition to providing a check that the product design meets customer requirements, the Design Validation Stage is used for producing benchmark results. As the product being designed is likely to be revised and refined over time, it may be necessary to demonstrate in the future that the improved product does not compromise existing capabilities. In other words, future design changes may have unanticipated, undesirable side effects. In this case, one would like to revalidate the existing capabilities (comparing to the benchmarks) in addition to validating the new improvements. The results of the Design Validation Stage should be used to prepare for this future activity.

If the production test model is tested in the end-user environment, and that testing is documented and controlled, then these tests could be rerun to provide the revalidation needed at a later time. In the author's experience with ISO 9001 auditing of engineering computer software, this revalidation activity has been deemed an ISO 9001 requirement by auditors. Before revised in-house engineering computer software could be released to users (even though it had already been approved for analytical correctness in a design verification procedure) the auditors said the software must then be compared to existing benchmark results, and a search for inadvertent compromises of the results must be undertaken.[1]

Whether or not this is an ISO 9001 requirement for all types of products (besides engineering computer software) is not certain, but it is clearly desirable to do checking for inadvertent adverse effects on revised designs and product enhancements. Assuming that information created by testing of production test models at the Design Validation Stage is properly maintained, the benchmark results needed for this checking should be readily available. If necessary, a separate report or form could be created to specifically record product performance or other needed benchmark data. As usual, if this documentation is included in the Development Project File, it should be easily referenced in an ISO 9001 audit situation.

Conclusion

This chapter has presented a discussion of the Design Validation Stage as envisioned by the author as part of an ideal ISO 9001 design and development process. The Design Validation Stage is assumed to be the final activity prior to the release of the product for final production runs and release to the customer. The basic aim of the Design Validation Stage is the check that the product design meets customer requirements. This can be based on a review of the customer requirements and the output from the Design Output Stage (final drawings, BOMs, etc.). It can also be based on testing and inspection of a preproduction or production test model of the product. Regardless of the method used, it must be documented, and the documentation must be controlled for use in the ISO 9001 audit situation. This documentation can be based on a standard form indicating that a review activity has taken place. The documentation of validation inspection and testing can be based on the engineering test sample documentation presented in Chapter 5.

Finally, this chapter on the Design Validation Stage and the preceding chapters [covering the Design Input Stage (Chapter 4), the Design Stage (Chapter 5), and the Design Output Stage (Chapter 6)] have presented details about the basic design and development aspects of the ISO 9001 standard by expanding upon the ideal ISO 9001 process. These details are intended to help the reader in adapting existing design and development processes to meet ISO 9001 requirements.

Although the detailed procedures, forms, reports, etc. may be practical for use in some organizations, it is unlikely that all of them can be used directly. These samples should also not be assumed to be guaranteed to be successful in an audit situation since they have been developed primarily as a tool to understand the ISO 9001 process requirements. A list of the standard tools and documents is found in Appendix A along with the appropriate concept or section of the ISO 9001 standard.

The remaining chapters in this book deal with understanding the ISO 9000 philosophy generally and the ISO 9001 standard specifically without regard to a specific set of sample procedures. Although the discussion in the subsequent chapters is tailored to design and engineering issues, the material covered in these chapters may only be of interest to engineering managers or designers and engineers looking to achieve a deeper understanding of ISO 9001.

References

1. A Non-Compliance Note from the author's experience with a Lloyd's Register Quality Assurance Limited ISO 9001 registration audit in 1993 stated the following: "Although appropriate steps were being taken to verify programs prior to use, when programs were amended records did not always demonstrate that all amendments were effective and had not compromised other, previously acceptable functions of the program." The section of the standard cited for noncompliance was ISO 9001-1987 paragraph 4.11 (Inspection, Measuring, and Test Equipment).

8

An Introduction
to Quality Systems

Introduction

Most engineers and designers do not have a formal background in the field of quality control. Although this book is not a reference for this field of study, there are aspects of ISO 9000 generally and the ISO 9001 standard specifically which are taken from this field. Some of these aspects should be understood by engineers and designers learning about ISO 9000.

This chapter begins an exploration of the quality field as seen in the ISO 9000 philosophy, starting with probably the most important aspect which should be understood—the quality system. It is hoped that most readers are going to find the information simple and obvious since there is no high degree of technical sophistication in these basic concepts (although parts of the quality control field such as statistical analysis and design of experiments can be rather sophisticated). Instead, the designer or engineer should try to become comfortable with the terminology and try to grasp the connection between the concepts and his or her own organization.

Quality System Definition

A central concept in the ISO 9000 series of standards is quality system. One definition of a quality system would be the following:

- *Quality system.* The combination of an organization's structure, tools, and written procedures which embody the means of achieving quality.

For the design and engineering organization, achieving quality could be clarified to mean creating products that consistently satisfy a customer's needs, as well as meeting the organization's needs.

For the design and engineering organization, structure, tools, and procedures have a very broad scope. The broad scope of design and engineering quality systems is evident when one considers that creating products that consistently satisfy customer's needs can involve entire companies, and those companies can use tools for such diverse activities as market research, contract specification, design, procurement, manufacturing, verification of incoming material, distribution, installation, and field technical support. Furthermore, for each of these (and probably other) aspects of product creation, qualified and well-managed personnel must be available to use the tools and carry out the organization's written procedures for customer satisfaction. The quality system as defined, then, practically involves understanding the whole functioning of an organization or company.

The definition of quality system may in fact appear too open-ended. There are so many activities that can have a direct impact on the ability of a product to meet the customer's needs, one could decide that everyone and everything in an organization is the quality system. Basically, there is no way to avoid this problem since ISO 9000 is supposed to work for any and all organizations (service as well as manufacturing, software as well as hardware, etc.). Whether this is prudent or not, it is clear that ISO 9001 is based on a commitment to completely universal application.[1]

The scope and level of detail required to constitute a quality system may remain uncertain until a given organization selects a specific standard from the ISO 9000 series of standards and selects a specific auditing firm. And, although the level of detail of one's quality system is always going to seem open-ended (and probably dynamic), it must be understood that a specific standard within the ISO 9000 series (such as ISO 9001) is going to dictate some minimum requirements for the quality system. These minimum requirements are discussed in subsequent chapters that discuss the ISO 9001 quality system.

Quality System as Documentation

Having presented a definition of quality system, we can consider a more pragmatic understanding of the quality system concept. In a practical sense, the quality system of an organization comes down to a set of documentation. Although the definition of quality system includes an organization's structure, the organizational structure is typically documented in some fashion. Although chains of command and organizational charts are not the only means of organizing personnel, it is the most common. Therefore, the organizational structure part of a quality system is likely to consist of documents which show the structure, relationships, and responsibilities among personnel.

Although qualified personnel are supposed to fill the positions defined in the structure, most likely this would also be demonstrated by documentation. For instance, personnel records are probably kept for each member of the organization which would indicate that a person has the proper credentials for filling a position. Again, the quality system concept appears to be distilled down to forms of documentation.

Although the tools (hardware as well as software) that are used by personnel must be capable of meeting the demands of the quality system, once again it is basically documentation that indicates this. If one is interested in the capability of a tool, one can create test plans and reports, one can do calibration reports, one can perform process capability studies on it. This documentation then also becomes a component of the quality system.

Finally, it should be obvious that the written procedures (as indicated by the definition of quality system) are documents as well. The degree of detail in written procedures may vary widely between products and organizations. However, it is assumed that there cannot be a formal quality system that can be understood, refined, or audited unless documentation which indicates the basic operation of the organization exists.

Thus the quality system, in a practical sense, can be thought of as a set of various kinds of documentation. Note that documentation is a valuable by-product of the quality system being considered. Specifically, the quality system has a paper trail. An organization can now be audited for actually having and/or using a quality system. The ability to perform such auditing (both internally and by third parties) is an essential ingredient in the success of the ISO 9000 series of standards.

Quality System as Management Model

Another facet of the concept of a quality system would be to consider it a management model. An organization's structure, tools, and procedures (the stated parts of a quality system) should, in fact, form a strong foundation of understanding how an organization functions and how it is managed. The quality system should be able to indicate the mission of each person or group; it should indicate the internal customers and suppliers of each person or group. Theoretically given this information, each person should then be able to perform error-free or quality work by meeting their internal customer's needs. With each person providing quality work, the overall output of the organization should be meeting the external customer's needs (assuming the quality system is the correct system). This is what one would expect of an effective management model—it gets the task done right, on the first attempt.

There are other advantages to considering the quality system to be a management model. First, it helps to make quality work a highly visible priority for levels of management. Second, it means that one can apply quality improvement ideas to management. Considering the quality system to be a management model implies that the effectiveness of managers can be measured, just as one can measure the effectiveness of tools and processes and products. If the performance of management can be measured, we can assume that it can be improved.

Theoretically, then, when problems occur in an organization, and they are then exposed to a quality-sensitive management, then the problem is going to be corrected. Furthermore, by improving the documented procedures, the problem is not expected to occur again. This concept gets to the heart of the ISO 9000 philosophy. If the quality system is documented and understood, and if one can then tell whether it is working properly or not, an organization should be able to consistently meet needs and continually improve itself. This concept is often referred to as "process improvement" or "continuous improvement."

Of course, in the ISO 9000 philosophy, it does appear that improvement would only occur when there are problems. Therefore, the latest versions of ISO 9000 series of standards are placing greater emphasis on proactive or preventative action. This is an important consideration for engineers and designers since this drive for preventative action can be used to demand that more care be taken during the design and development stages. Most managers in the author's personal experience seem to feel that designers and engineers take too long; they always want to keep analyzing designs. However, if the management really wants preventative action, then there would be no better place to take more time to "get it right" than in the design phase, particularly since it is easiest to effect real change in the product during this phase.

Another advantage of seeing the quality system as a management model is being able to more easily incorporate other quality-oriented programs such as Total Quality Management (TQM). A TQM program encourages members of an organization to strive for such goals as zero defects (in processes such as design and development as well as zero defects in manufactured products), and managing by preventative action (working to prevent quality problems instead of correcting problems after they appear).

These TQM-type goals, at least from the management's point of view anyway, may be achieved by enhancing or refining a quality system which also meets the demands of ISO 9000. Unfortunately, more than just changing written procedures is going to be needed to achieve error-free work in the design and development environment. In many cases, the tools and personnel involved in design and engineering are just as important or more important than a set of written procedures, particularly since the art and science of design and engineering is changing so rapidly. Furthermore, state-of-the-art design work that probes the limits of feasibility in the physical universe must involve failures as well as successes.

Quality System in Practice

Before showing some examples of quality systems, Figure 8.1 presents the definition of quality system, as well as some practical considerations of its working definition. With respect to ISO 9001, although the scope and detail of a quality system may be subject to

---

QUALITY SYSTEM

Definition:

The combination of an organization's structure, tools, and written procedures which embody the means of achieving quality.

Features:

❑ **It is made up of documentation of various kinds**

❑ **It includes the organizational structure**

❑ **It includes the organization's most basic practices**

❑ **It specifies the organization's quality tools such as auditing**

❑ **Its top "level" of the documentation is the Quality Manual**

❑ **It can have a very broad scope for design and engineering**

Figure 8.1 Quality system.

interpretation based on the type of organization, ISO 9000 standards (such as ISO 9001) demand that certain components be found in the quality system.

Examples

Next we consider some examples of quality system components. The first example deals with organizational structure. The second example presents organizational tools. The last example deals with organizational documentation.

Organizational structure

Most organizations have hierarchy-based structures. The levels and precedence are shown according to organizational charts. The traditional charts show each level of management or each department. For each level or department, there is a position of authority with lower personnel reporting to the person in the position of authority. In some cases, there may be parallel or multiple lines where people in those positions report to a variety of departments. Figure 8.2 shows some typical organizational charts.

It must be pointed out that the quality system (at least one based on ISO 9000) probably demands a certain amount of top-down-style management. This is because the ISO 9000 quality system must define a top level (or executive) level of management that can dictate to lower levels of management. This arrangement allows the organization to show that there is a position of authority with control over the quality system (and that this organizational model theoretically makes sure that quality problems, permanent corrective action, and preventative action are addressed in a timely manner). Thus, although TQM attitudes such as "quality is everyone's responsibility" are desirable, for ISO 9000, the organization must still show a level of management with a final say in matters of quality. This does not necessarily have to conflict with the TQM attitude, but in practice members of an organization may naturally revert to stating that ensuring quality is really the quality control manager's function.

In addition to using organizational charts to demonstrate the authority of the organizational structure, such charts can demonstrate lines of communication or technical interfaces. This can be useful in demonstrating an attempt to implement cross-functional operation. However, showing "dotted lines" between managers or groups does not necessarily demonstrate effective communications are occurring between functional areas that have a direct impact on the design process (an ISO 9001 requirement). Just the same, a set of well-main-

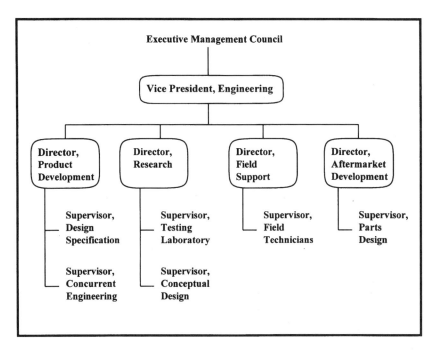

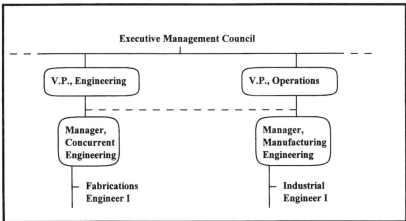

Figure 8.2   Organizational charts.

tained and well-understood organizational charts is an excellent practical example of the organization's structure as desired by the working definition of quality system.

Organizational tools

The next practical example of the quality system is its tools. An excellent example of a quality system tool is audits. In this case, the audit is considered an internal or first-party audit by means of which members of the organization perform a self-examination procedure. One should note that these kinds of audits and their procedure are a specified requirement for ISO 9001. Although any tool that is necessary for achieving quality can be considered part of the quality system, audits provide a most useful example of a quality system tool since every organization interested in the ISO 9000 series of standards is going to have to use this tool.

As a tool in an organization's quality system, audits must be specified by and conducted according to a documented procedure (as in the Quality Manual mentioned in Chapter 9). Furthermore, the organizational structure dictates what part of the organization is responsible for conducting these audits. The part of the organization responsible for the audit obviously needs some amount of independence, and this need is addressed in the ISO 9000 series of standards.

In order to conduct the audit, quality auditors might decide to study the documented procedures of a part of the organization (or department). After studying these procedures and probably discussing them with the management of the department, the auditors would then study and inspect the department. In the case of the design and engineering organization, this inspection would include questioning members of the department, studying the progress of a project, or studying documentary evidence of quality system performance.

For example, the auditor may question a designer to see if he or she knows where to find their controlled procedures, the auditor may check if the designer knows how to request corrective action, or the auditor may see if he or she knows the organization's most basic quality system statement—the Quality Policy. As another example, the auditor may question an engineer to see if he or she has performed verification calculations in conjunction with a product whose drawings are due to be released for manufacture. Instead of questioning members of the department, an auditor might also study drawings, calculation sheets, contract specifications, or minutes of design review meetings. In any of these scenarios, the auditor is looking for noncompliance with the documented quality system procedures. The auditor may also find procedures that lack components necessary for ISO 9000-based quality systems.

When the auditor discovers noncompliance items, a procedure to correct the problem must be put into motion. Furthermore, the auditor's part of the organization must verify that these items are being corrected within a reasonable amount of time. Clearly this process of constantly checking if the quality system is in place, and checking if it is being followed, is a tool. As we have seen, tools are an essential component of the quality system.

Organizational documentation

The last practical example of the quality system is an example of documentation. This example is the Quality Manual. The term "Quality Manual" is common in the area of quality management. In many cases (whether appropriate or not), the term "Quality Manual" is substituted for quality system. The Quality Manual is generally the documentation that embodies the quality system in use by an organization. The ISO 9004 guideline document mentions the Quality Manual in this light, and it makes reference to another ISO standard for more guidance on the subject of quality manuals.[2]

The Quality Manual would typically include the documentation specifically required by ISO 9000. This mandatory documentation includes the Quality Policy (which is discussed in Chapter 9). The Quality Manual would also include the organizational charts, management process flow diagrams, and definitions of responsibilities.

Depending on the size of the organization, the Quality Manual may be the only documentation needed for the quality system. However, in most organizations dealing with ISO 9001, there will be a hierarchy of documentation. The Quality Manual or other top-most level operating policy documentation will then refer to the lower-level documentation (such as engineering standards and departmental work procedures). These working procedures may then specify that documentation such as forms, checklists, reports, and meeting minutes are to be created and maintained. Although completed forms or checklists would not typically be part of the Quality Manual, the design or content of the forms can be part of the Quality Manual. The size and complexity of the Quality Manual is clearly going to be dictated by the complexity of the organization and its products.

It is essential that the reader understand that documentation such as the Quality Manual is to be considered special. It is to be considered controlled. This means that the documentation is automatically updated by the issuer of the document within the organization. Controlled documentation may not be given to anyone at any time; it cannot be revised or removed at random. This caveat is necessary since the quality system cannot succeed if it is not followed, and it cannot be followed if everyone in an organization does not have the proper documentation in its most recent version. Therefore, the Quality Manual (along with various other documentation meeting the requirements of ISO 9001) is to be a controlled document.

Note that the Quality Manual is the most visible and probably the most important example of documentation in the quality system. This documentation, along with an organization's structure and tools, makes up the working definition of quality system used in this book. There are many other examples of tools (engineering software, test equipment, and statistical methods, for instance), structure (product improvement teams and steering committees, for instance), and written documentation (drawings, bills, reports). However, it is hoped that the examples presented in this chapter have given the reader a basic and fair understanding of the quality system concept. The ISO 9001 requirements for a specific quality system (at least with respect to product design and development) are addressed in detail in Chapter 9.

Conclusion

This chapter has presented an introduction to the quality control aspects of the ISO 9000 series of standards by presenting a working definition for quality system. As the titles of the standards in the ISO 9000 series indicate,[3] quality system is a fundamental concept for standards in the series such as ISO 9001. Finally, the quality system concept can be seen in a practical sense as the functioning management of an organization (management model) and as its set of documented procedures and records.

References

1. According to the ASQC *On Q* journal in an article entitled "Committee Recommends RAB Software System Registration," edited by Penny Fredrick (February 1994, p. 11), a committee was formed by the Registrar Accreditation Board (RAB) to advise on the need for a U.S. accreditation program for the application of ISO 9001 to software. This Software Quality System Registration (SQSR) Committee's recommendation for a software-specific arrangement was rejected by the RAB (Information Technology Association of America correspondence dated August 25, 1994 from Douglas C. Jerger). Despite the distinct nature of software development, ISO 9001 is still expected to apply with its usual auditing procedures.
2. ISO 10013, *Guidelines for Developing Quality Manuals* is listed in the Bibliography of ISO 9004 (ANSI/ASQC Q9004-1994). Milwaukee, Wisconsin, 1994.
3. ISO 9001 (ANSI/ASQC Q9001-1994) is entitled *Quality Systems—Model for Quality Assurance in Design, Development, Production, Installation, and Servicing.* ISO 9002 (ANSI/ASQC Q9002-1994) is entitled *Quality Systems—Model for Quality Assurance in Production, Installation, and Servicing.* ISO 9003 (ANSI/ASQC Q9003-1994) is entitled *Quality Systems—Model for Quality Assurance in Final Inspection and Test.* Milwaukee, Wisconsin, 1994.

9

Quality System Guidelines

Introduction

Having developed a working definition of a quality system in Chapter 8, we need to develop a more complete view of the quality system. Such a view is based on a document called ISO 9004, which is a guideline document in the ISO 9000 series of documentation.[1] ISO 9004 is not a standard against which an organization would be audited.

Since a quality system that incorporates the ISO 9004 concepts can be considered a basic ISO 9000 quality system, this guideline is presented in detail in this book so that the reader can determine if the processes in his or her organization currently does, or could in the future, incorporate these concepts. To make the most of the examination of ISO 9004, a copy of the ISO 9004 document should be obtained and used as a reference while reading this material.

The understanding of this basic quality system is also essential to understanding the basic requirements of ISO 9001. The ISO 9001 standard is almost completely devoted to specifying the basic requirements of the quality system as opposed to demanding specific activities, so it is essential to have an understanding of the ISO 9000 basic quality system (as perceived in ISO 9004) before studying ISO 9001.

ISO 9004 Quality System Definition

ISO 9004 defines a quality system as "the organizational structure, procedures, and resources needed to implement quality management."[2] ISO 9004 states that quality management "encompasses all activities of the overall management function that determine the quality policy, objectives, and responsibilities, and implement them by means such as quality planning, quality control, quality assurance, and quality improvement within the quality system."[3] In other words, the quality system is expected to implement whatever an organization's management determines are appropriate "policies and objectives."[4]

For the design and engineering organization, a typical objective would be creating products that consistently meets customer's needs. A likely policy would include compliance with state and federal laws in various forms as well as complying with ISO 9001. Although ISO 9004 does not specify a policy or objective for management to achieve with the quality system, ISO 9004 does give some guidelines for quality policy and objectives.

Quality Policy

Paragraph 4.2 of ISO 9004 indicates that there should be a documented quality policy. This document is the central document for the entire quality system. It gives the topmost level of management a specific point of entry to the documented procedures (or Quality Manual). All other procedures exist to support the stated policy. Paragraph 4.2 further indicates that the policy should be consistent with other policies in the organization (such as following pertinent laws and regulations). Also, paragraph 4.2 indicates that management is expected to ensure that the quality policy is understood, implemented, and reviewed at all levels of the organization. This generally means that all the members of an organization could be asked to locate and demonstrate understanding of the quality policy (in an audit situation, for instance).

Quality Objectives

Paragraph 4.3 of ISO 9004 presents guidelines for quality objectives. Quality objectives could be targets for on-time shipments, or designs within schedule and budget, or a minimum of rework, for instance. Objectives such as these are to consider "fitness for use, performance, safety, and dependability."[5]

ISO 9004 also mentions a cost of quality approach under the quality objective paragraph (Paragraph 4.3.2). This involves tracking the costs associated with quality and minimizing any negative costs. For the design and engineering organization, this might mean establishing a measurement (or metric) of how often manufacturing needs to request a change in drawings due to drafting errors. The measurement could just consist of the number of drawing revisions of this type that must be done per month. However, ISO 9004 suggests that a better measurement might be the amount of time spent finding and correcting the error times an hourly charge rate. This metric would track the cost of quality (COQ) and could be a more meaningful measurement. Finally, as a quality objective which meets the ISO 9004 guidelines, the design and engineering organization would set a minimum or target for this metric.

Quality System Elements

As shown by ISO 9004, the quality system exists to meet the quality policy and thus the objectives of the organization. However, ISO 9004 does further present some guidelines for what specific features and attributes should be found in a quality system. This material starts with paragraph 5.

After presenting generic quality system information in paragraph 5, ISO 9004 goes on to present information in specific functions such as finance, marketing, design, purchasing, production, testing, and safety. The remainder of this chapter focuses on the fundamental information which all functions or groups in an organization should understand (including design and engineering) and the specific information which relates to the design function.

Life cycle

Paragraph 5.1.1 of ISO 9004 states that the quality system should cover all parts of the life cycle of a product. This includes market research through product disposal (Figure 9.1). Although some organizations may only be concerned with parts of the life cycle, a design and engineering organization seeking ISO 9001 compliance is likely to be concerned with all the phases in the cycle.

Furthermore, the design and engineering organization should be aware that the next paragraph 5.1.2, gives special consideration to the design and associated phases. According to this paragraph "within an organization, marketing and design should be emphasized as especially important for...determining and defining customer needs, expectations, and other product requirements, and...providing the concepts (including supporting data) for producing a product to documented specifications at optimum cost."[6] It appears from this statement that part of the philosophy behind the ISO 9000 standards is an elevated importance for the design function and customer interface functions.

Management

As mentioned in Chapter 8 introducing quality systems, the ISO 9000 system basically seems to assume a traditional, top-down, management model. The top-level or executive management sets the policy, and progressively lower levels supervise and perform activities that

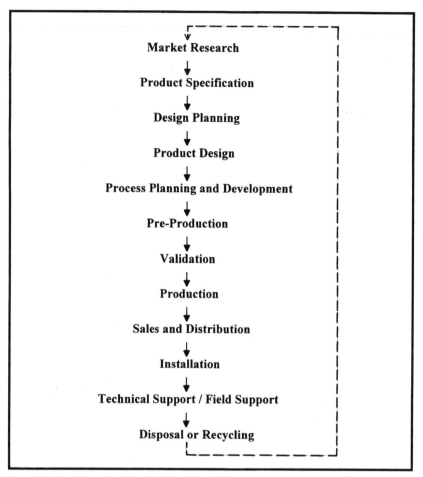

Figure 9.1 Sample product life cycle adapted from ISO 9004.

implement these policies. Regardless of whether organizations (and design and engineering types of organizations in particular) actually behave in this manner, ISO 9000 presumes that this is the prevailing management system. In keeping with this philosophy, ISO 9004 specifies some management guidelines.

In paragraph 5.2.1, ISO 9004 states that "input from the market should be used to improve new and existing products and to improve the quality system. Management is ultimately responsible for establishing the quality policy and for decisions concerning the initiation, development, implementation, and maintenance of the quality system."[7] Considering the broad scope of the quality system, this statement essentially seems to be stating that management is to be held responsible for the success or failure of the organization.

In paragraph 5.2.2, ISO 9004 seems to present the need for a management model within an organization. This paragraph states that someone (presumed to be management) must determine the "activities that contribute to quality, whether directly or indirectly."[8] These activities are then to be defined and documented. Considering the scope of the ISO 9004 life cycle and the qualifier of contributing to quality directly or indirectly, it appears that everyone in the organization is expected to have a defined position within the organization's structure. This is certainly the case for the design and engineering organization. Anyone involved with the design activity (drafters, designers, engineers, data entry clerks, BOM processors, administration personnel, etc.) are going to directly or indirectly contribute to quality. Therefore, their position within the organizational structure needs to be defined.

Paragraph 5.2.2 goes on to state that each defined position should have "general and specific quality-related responsibilities"[9] defined; that "responsibility and authority delegated to each activity contributing to quality should be clearly established"[10]; and that "interface control and coordination measures between different activities should be defined."[11] Thus, for each position in the organization, there should exist job descriptions, job qualifications, clear lines of communication, and authority. Furthermore, each of these items would need to be documented, and these documents would need to be controlled to be certain they are current and accurate.

Job descriptions and qualifications would typically be personnel information kept by a human resources department or supervisors. The lines of communication and authority would typically be found in organizational charts controlled by the management of the organization. This aspect of the ISO 9000 quality system should not be difficult to satisfy since this type of information most likely already exists within organizations.

Paragraph 5.2.2 also indicates that attention should be given to allowing the freedom to prevent and correct problems. This paragraph includes the statement "organizational freedom, and authority to act should be sufficient to attain the assigned quality objectives with the desired efficiency."[12] In most cases, this statement means the documented management model needs to document that each member of the organization has the right to bring attention to problems within and beyond their department (or at least to initiate a formal request for corrective action). In a sense, this relaxes somewhat the top-down management philosophy of ISO 9000. Although management is responsible for setting the quality policy and objectives, lower level members are encouraged to indicate when the policy and objectives are not being met.

Although it is implied by paragraph 5.2.2, paragraph 5.2.3 states that the quality system activities (i.e., activities affecting quality such as design) are to be organized into a structure. These quality-related activities are to be "clearly established within the overall organizational structure."[13] In the design and engineering organization, it is suggested to consider all activities as quality-related. Then the existing management model (lines of authority, job definitions, etc.) for this organization can just be expected to fulfill the need for organizational structure.

In addition to giving structure to the entire organization, the established organizational structure should contain a quality function. In this case, paragraph 5.2.3 could be interpreted as indicating that "functions related to the quality system"[14] means maintaining quality records and manuals, holding internal audits, and verifying corrective and preventative action. In most organizations, these functions are performed by a quality assurance or quality control department within the organization. Although ISO 9004 does not specify this activity specifically, ISO 9001 does make this clear. ISO 9001 also specifies that there must be someone at the executive management level responsible for the quality system.

The next aspect of the management guidelines in ISO 9004 concerns the allocation of resources and personnel. This allocation reinforces a traditional management model. That is, management is responsible for setting the policy and objectives and for achieving the objectives. Paragraph 5.2.4 states that management should "provide sufficient and appropriate resources essential to the implementation of the quality policy and achievement of quality objectives."[15] Therefore, a quality system is not necessarily viable unless it can be practically implemented given actual resources within the organization. As an example, an organization should not dictate that historical records be kept in an electronic format and then fail to acquire proper computer equipment. As another example, management should not determine that process capability statistical data be kept for a work area, but never hire someone with the skill to implement this procedure.

The need to obtain proper resources to make the quality system work is applied rather broadly. Management's responsibility as seen in paragraph 5.2.4 includes human resources, development equipment, manufacturing equipment, test equipment, instrumentation. It could easily include an appropriate, secure, and safe work environment. In terms of human resources, management is expected to determine the "level of competence, experience, and training necessary to ensure capability of personnel."[16]

Operational procedures

With paragraph 5.2.5, ISO 9004 begins to define the quality system as documentation. This paragraph starts by saying the "quality system should be organized in such a way that adequate and continuous control is exercised over all activities affecting quality."[17] Shortly it will be seen that this control is supposed to result from documenting proper procedures and then assuming that they will be followed by members of the organization. After suggesting that the quality system should emphasize preventative actions in addition to corrective action, this paragraph states that "documented operational procedures coordinating different activities with respect to an effective quality system should be developed, issued, and maintained to implement the quality policy and objectives."[18] Further on, this paragraph states "documented procedures should be stated simply, unambiguously, and understandably, and should indicate methods to be used and criteria to be satisfied."[19]

As discussed in Chapter 8, the quality system can be thought of as being encapsulated by the organization's quality documentation. This documentation is often referred to in the Quality Manual. This quality system documentation is critical to successfully conforming with any ISO 9000 standard.

Configuration management

The next paragraph in ISO 9004 concerns configuration management. Configuration management is concerned with the identification and status of products or services in their life cycle. This identification and status lets suppliers and customers (internal and external) determine the state of a product and its documentation. It should be clear that the quality system cannot function if it cannot distinguish between different product types, or if it cannot determine whether a product is in development or production or in some other stage of its life cycle.

Configuration management has always been a critical issue in the design and engineering organization, and it should not be difficult for most of these organizations to comply with the requirements. Most design and engineering organizations have a system of identifying different products through model names, model numbers, etc., and the vintage of products through serial numbers, production control systems, etc. Beyond this system, there should also be in place a way of distinguishing products that are in a development stage versus production. Of course, items such as drawings, engineering reports, and service manuals are to have revision levels. Finally, there should be master documentation for the models, drawings, reports, and serial information. This master documentation should permit someone (such as an auditor) to determine the stage in the life cycle for any given product, what documentation (drawings, manuals, etc.) apply to that product, and any important historical information for that product (such as safety bulletins).

Quality system documentation

As mentioned earlier, an organization's quality system is strongly tied to the organization's documentation. Although any combination of organizational structure, tools, and procedures can be considered a quality system, it cannot meet ISO 9000 requirements until it is documented and approved by the organization's management. Once the quality system is written down and it is available to members of the organization, it can be implemented and it can be determined whether the implementation of such a system is actually achieving quality. Thus, documentation is crucial to quality systems, and documentation is crucial in any ISO 9000 investigation, audit, or registration.

One crucial aspect of the quality system discussion is deciding how much documentation to consider part of the quality system. It has already been mentioned that documentation of the quality system must include the Quality Policy and the relevant organizational structure (charts, lines of authority, and communication). As mentioned in Chapter 8, these static and relatively simple documents are usually contained in a document called the Quality Manual.

The handling of quality system documentation becomes more uncertain with respect to procedures and records. As mentioned in the previous section, configuration management documentation is supposed to be maintained by an organization. Does this catalog type of information belong in the Quality Manual itself, or should it be maintained in a separate document whose structure is defined in the Quality Manual? This question is not clearly answered in the ISO 9000 guideline. In a large organization, the configuration management would probably constitute a separate document system. However, in a small organization where the Quality Manual may be the only controlled document, new product identifications might be included in the Quality Manual.

The flexibility in scope and size of the quality system documentation is mentioned a number of times in the ISO 9004 guideline. According to paragraph 5.3.1, "care should be taken to limit documentation to the extent pertinent to the application."[20] And in paragraph 5.3.2.4, it is stated that "documented procedures can take various forms, depending on the size of the organization, the specific nature of the activity, and the intended scope and structure of the quality manual. Documented procedures may apply to one or more parts of the organization."[21]

Although the quality system documentation is going to be as varied as the organizations that develop it, some basic documentation rules can be followed. In the configuration management example mentioned earlier, a large organization would probably have the structure of the configuration management defined in the Quality Manual. However, the configuration management documentation itself would not be in the Quality Manual. Thus, there is a natural distinction between documentation definition and its resulting records.

A common model of defining types of documentation refers to levels or tiers of documentation (Figure 9.2). The top level can be seen as the Quality Manual. The Quality Manual contains the organization's quality policy and its organizational structure. The Quality Manual is probably the province of the top level of management. The Quality Manual serves to define the different departments and groups within the organization.

In addition to defining the partitions of the organization, the Quality Manual needs to reference the procedures used by these partitions. These documented procedures should be considered part of the quality system, but they would form a larger body of documentation in a middle level. Paragraph 5.3.2.4 mentions "design, purchasing, and process work instructions"[22] as examples of areas of documented quality system procedures that support the Quality Manual. These departmental procedures would be the responsibility of middle levels of management since they would be documenting processes and resources used for their respective departments. For an industrial type of organization, this means that the engineering and manufacturing disciplines would be able to develop their own organizational procedures that best fit their activities.

The bottom level of documentation can be considered quality records. These are not procedures, but documents dictated by procedures. For instance, an engineering department procedure may dictate that a meeting be held between designers and marketing representatives before a set of drawings is released for manufacture. The procedure (from the middle level) should dictate that a document be prepared and signed off to signify the design review meeting was held. That document can be considered a quality record. It is clearly not part of the Quality Manual or the department procedure; thus it can be considered as part of the third or bottom level of documentation.

Although at the bottom level, quality records are by no means unimportant. This type of documentation is evidence that the quality system is actually being followed, and this is obviously crucial in an

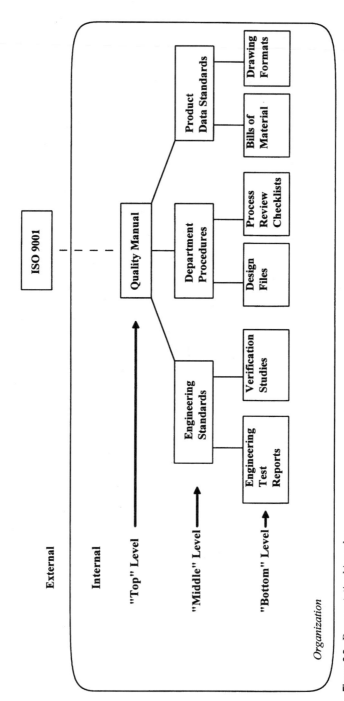

Figure 9.2 Documentation hierarchy.

audit situation. Thus, the quality records need to be maintained and controlled so that they are in order and accessible at the time of an internal or external audit. Indeed, one should be aware that all three levels of documentation must be controlled and maintained.

Although it is important to create workable quality system documentation, it is equally important to maintain this documentation. In the ISO 9000 quality system, the activity of maintaining documents is often referred to as document control, and there is a recognized distinction between documents that are controlled versus those that are uncontrolled. A basis for this distinction can be found in the ISO 9004 guideline. Paragraph 5.3.1 states that the "quality system should include adequate provision for the proper identification, distribution, collection, and maintenance of all quality system documents."[23] Also, paragraph 5.3.2.3 states, "Documented procedures should be established for making changes, modifications, revisions, or additions to the contents of a quality manual."[24] Therefore, there must be a system in place to see that the proper documentation is available at the proper revision level to the proper members of the organization. This system is to be documented in procedures that are part of the top level of documentation. Finally, note that a document control system is to apply to all quality system documentation (i.e., all levels).

As an example, consider the Quality Manual itself. This top-level document is going to contain at least the Quality Policy, organizational structure documentation, and now also the documentation control system. This Quality Manual needs to be readily available to all levels of management in the organization. This is considered important since managers who are developing departmental procedures are to verify that their procedures are in accordance with the overall organization's policy and objectives. Furthermore, all levels of the organization (including nonsupervisory personnel) are supposed to understand the quality policy (ISO 9004 paragraph 5.2). Therefore, this group must also be able to access the Quality Manual (although probably not as often as managers). Given these conditions, how should the organization provide access to the Quality Manual?

A common solution in this case is to issue a controlled, numbered set of copies of the Quality Manual. Usually, a specified department in the organization is responsible for duplicating, numbering, and distributing these copies. Furthermore, this department is responsible for maintaining a master list of the copy numbers and the persons receiving the document. In the case of the Quality Manual, a copy could go to each manager of each department in the organization. In addition, some managers may be issued second copies for access by a wider audience. These copies might be placed on a shop floor, or in a drafting area, or in a library. Nonsupervisory personnel would then be informed that these copies are for their use.

The crucial aspect of this system is that if revisions are required for the Quality Manual, the responsible department can use its master list to issue a new copy of the Quality Manual (or part of it) to the holders of the document. It is this ability to automatically update holders of the document and ensure that the latest version is available that makes the Quality Manual (or any other documentation treated in this manner) controlled. Note that this department would likely also be the focal point for requesting revisions, editing documents, getting revisions approved, etc.

Considering current technology, it should be mentioned that electronic media can be used to achieve document control. ISO 9004 paragraph 17.3 states that "Records may be in the form of any type of media such as hard copy, electronic media, etc."[25] So there should be no problem in using computer-based systems.

For example, a computer network could be established with a file server that has the one copy of the Quality Manual on-line. Users on the network would then be able to view this one copy of the Quality Manual. The network would be defined so that only an approved member or members would be able to change the one copy of the Quality Manual (Figure 9.3 shows an arrangement for a typical electronic documentation system). Regardless of which system of control is used (printed or electronic media), the procedure must be documented and must be effective at ensuring that members of the organization have the proper documentation at all times.

Inevitably, members of the organization may desire or need copies of the controlled documents. If members make these copies, such copies are considered uncontrolled (since the person making the copy will not be automatically informed of updates to the controlled document). Of course, the problem arises that someone may mistake them for being controlled documents. There are a few solutions to this problem. One solution is never to make uncontrolled copies. Despite the fact that this policy could be extremely inconvenient, the organization could dictate such a policy, and it may be the best solution. Another solution is to clearly identify the copies as uncontrolled. This can be accomplished by having members of the organization write or stamp the word UNCONTROLLED on the copies. Doing so would inform anyone who used the copy that it may not represent the latest revision of the document, and that they should then be required to check its disposition.

The situation involving uncontrolled documents may not be a problem with somewhat static documents such as the Quality Manual. However, document control is to be used for all quality system documents (proce-

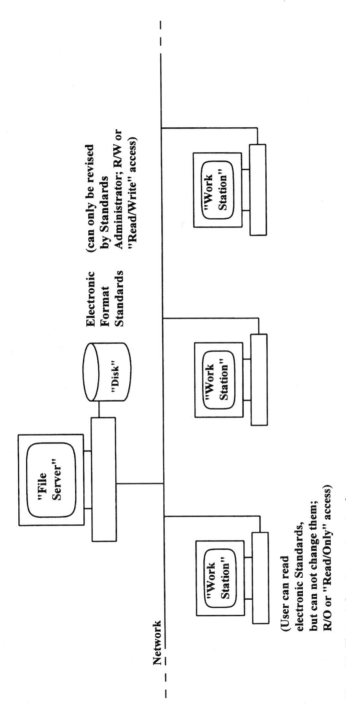

Figure 9.3 Electronic format computer network.

163

dures, quality records, etc.). In fact, documents such as department procedures are going to be more dynamic and more likely to be used on a daily basis. An organization's management must then decide how to handle the situation whereby personnel are constantly accessing a shared hard copy for information on how to perform their work. Another problem, of course, may be convincing personnel to follow through with changes in work processes when the written procedures are changed.

Typically, in the design and engineering organization, there may be another class of documentation to manage. This class of documentation is known as reference. Although a designer may reference controlled codes and standards (such as the ASME Pressure Vessel Code), he or she may also reference textbooks or technical papers that have no means of control. Such documentation can be considered reference material. Although copies of this documentation might be marked or stamped FOR REFERENCE ONLY, the organization might decide it is impractical to mark all such copies in the entire organization. In this case, the organization may decide it is easier to mark controlled documentation as CONTROLLED and then have a policy that anything which is not marked controlled or uncontrolled is to be considered reference documentation. It should be noted, however, that if an equation or procedure taken from reference documents is being used frequently, then the equation or procedure should be incorporated into a controlled engineering standard or department procedure.

Quality plans

The next subject covered in the ISO 9004 quality system guideline document is quality plans. Quality plans can be seen in two different contexts. The first is a product context; the second is a process context. In the product context, a quality plan documents the quality objectives to be attained for that product. For the design and engineering organization, this product quality plan is essential for establishing the expectations of the design output. In the process context, a quality plan documents the procedures used to achieve quality objectives. For the design and engineering organization, then, this process quality plan constitutes the standard procedure that is to be used to create quality designs.

The dual nature of the term "quality plan" in the ISO 9004 guideline is seen directly in paragraph 5.3.3: "For any product or process, management should ensure that documented quality plans are prepared and maintained. These should...ensure that specified requirements for a product, project, or contract are met."[26] Notice that this reference starts by stating product or process.

Further on, the product context of quality plan is seen in the following statement: "Quality plans should define: (a) the quality objectives to be attained (e.g. characteristics, uniformity, effectiveness, aesthetics...)."[27] Thus, quality plans should be prepared and considered controlled documentation for the product. The next item listed states that quality plans should define "(b) the steps in the processes that constitute the operating practice of the organization."[28] Thus, quality plans should be prepared and considered controlled documentation for the procedures used in the organization.

Although it may seem helpful to lump product and process in many groups of the organization, for the design and engineering group they should be thought of separately. The product quality plan is going to change rather often with each new product design (and this quality plan would be associated with that design). The product quality plan may also differ from one section of a design and engineering group to another section since the group may be designing more that one type of product at a time. On the other hand, the process quality plan is most likely to be more generic. The process quality plan is associated with the organization or its groups. Although some design and engineering organizations may try to develop a process quality plan for each new product designed, most organizations are going to attempt to perfect a standard design practice and apply it as often as possible.

Product quality plan. For the design and engineering organization, the product Quality Plan may appear in different ways. If such an organization is going to develop a brand new product or design, the quality plan would be a plan for developing that product. This plan would include the quality objectives for the new product, but this can easily be in the process of change as the limitations or opportunities present themselves in the development process. When this development project quality plan is prepared, it should consider not only what attributes are needed in the new product, but it should also consider as soon as possible the safety issues, the market research issues, and the verification and validation issues.

If a design and engineering organization develops basically the same product for varying customer requirements, then the product quality plan is likely to appear as the customer specifications. In this case, it is important to document the customer's requirements and then specify the existing procedures and processes that are needed to meet these requirements. The quality plan for a standard product should also show documented consideration for safety and regulatory issues, procedures for revising the customer specifications, and how the design is going to be verified, validated, inspected, or tested.

Process quality plan. Looking back at the definition of a quality system, recall that meeting customer's expectations as defined by the product quality plan is not necessarily sufficient. It is equally important to provide the proper tools and procedures. This is supposed to be accomplished using a process Quality Plan. The process quality plan should generally be able to be specified in departmental procedures (or middle-level documents). The departmental procedure, in fact, should generally consist of what is mentioned in ISO 9004 paragraph 5.3.3.; namely, "the steps in the processes that constitute the operating practice of the organization."[29]

Paragraph 5.3.3 also mentions an excellent tool for understanding and documenting a department's operating practice—the flowchart. According to this paragraph, "a flowchart or similar diagram can be used to demonstrate the elements of the process."[30] Besides being an excellent way to demonstrate a documented process quality plan, it is an excellent tool for determining what process is actually used in a group or department. If one cannot create a flowchart for the processes in a department, most likely one does not really know how work is being accomplished by that department.

As an example, a department may be responsible for doing stress analysis for various components as a service to a larger engineering organization. This department would probably follow a general procedure for determining the work to be done, defining a job number or order identification, scheduling the work, performing the work, checking or verifying the results, validating that the results meet the requirements, and closing the job and providing technical support for the final results (Figure 9.4). A flowchart is an excellent way to document this process. Not only does it show the elements of the process, it indicates a direction or flow of the work. A flowchart can also show decision or acceptance criteria.

Audits

The next component to be found in an ISO 9000 quality system is the systematic application of audits. ISO 9004 states that "audits should be planned and carried out to determine if the activities and related results of the organization's quality system comply with planned arrangements, and to determine the effectiveness of the quality system."[31] Thus, the auditing is supposed to determine if the quality system is functional, if it is being followed, if it is effective in achieving quality.

There are different kinds of audits to be considered. In the case of the "internal audit" (sometimes called the first-party audit), an independent group within an organization audits (or inspects) other depart-

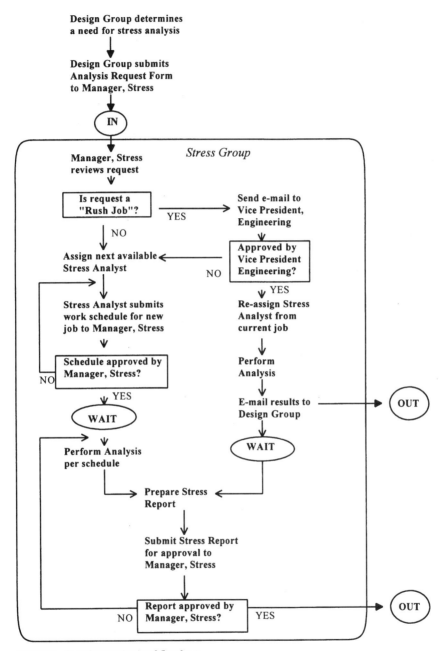

Figure 9.4   Sample organizational flowchart.

ments. Paragraph 5.4.1 states that "an appropriate audit program should be established and implemented by the organization's management."[32] Typically, the internal audits are conducted by the quality assurance or quality control department within the organization.

The quality group's audit personnel are to have "appropriate qualifications to conduct audits"[33] and "should be independent of those having direct responsibility for the specific activities or areas being audited."[34] The audit personnel would schedule an audit for each department involved with the quality system (and certainly the design and engineering group). During the audit, the auditor would review department procedures looking for the process quality plan and quality records that are specified. The auditor may then observe or question the personnel to see if the process quality plan is being followed, or the auditor may study the quality records to see if they are in compliance with documented procedures. The auditor may also question if members of the department are aware of the quality policy.

If an audit discovers a problem (procedures inadequate, procedures not being followed, tools unavailable or not capable, etc.), the organization is to establish a system of documenting and tracking these noncompliances. The organization is to have a system in place to see that the problems are corrected. The status reports of the results of the audit system must also be reported to the executive or top-level management. As stated in paragraph 5.4.5, "implementation and effectiveness of corrective actions resulting from previous audits should be assessed and documented."[35]

Another type of audit is called the second-party audit. In this case, members of one organization audit another organization. Typically, this involves a customer visiting a vendor's or supplier's organization to see that incoming product is under proper control.

The final type of audit considered is the third-party audit. This is the type of audit most people think of when discussing audits (at least for ISO 9000). In this case, a qualified and completely independent organization is contracted to evaluate a quality system. Examples would be an ISO 9001 audit (performed by a registrar) or a government agency audit.

In paragraph 5.5 of ISO 9004, it is stated that an "organization's management should provide for independent review and evaluation of the quality system at defined intervals."[36] These audits will not be able to cover the entire organization. The auditor may be visiting a large facility and may only have a few days to prepare the audit. There would be no way to cover the entire organization. As a consequence, the third-party auditor must rely on the results and effectiveness of the internal audit system. The third-party auditor may be looking at problems discovered during internal audits, and then verifying that the problems have been solved or in process of being solved. In this fashion, the auditor "leverages" the efforts of the internal audit system.

Quality improvement

Another attribute of quality systems discussed by the ISO 9004 guideline is quality improvement. Initially, meeting the ISO 9004 quality system requirements only produces a quality system that shows that the organization follows its own procedures. However, the procedures should not just be followed, but improved. This means that the quality system is to help advance quality improvement. Although documented procedures cannot dictate creativity in solving an organization's problems, it can encourage problem solving. The documented procedures can also be a vehicle for documenting and tracking process quality improvements. This means that improvements in one part of the organization can be transplanted to other parts by forwarding and adapting the documented procedures.

Paragraph 5.6 of ISO 9004 covers quality improvement. This paragraph says "the management of an organization should ensure that the system will facilitate and promote continuous quality improvement"[37] and it states that:

consideration should be given to:

a)   encouraging and sustaining a supportive style of management;

b)   promoting values, attributes, and behavior that foster improvement;

c)   setting clear quality-improvement goals;

d)   encouraging effective communication and teamwork;

e)   recognizing successes and achievements;

f)   training and educating for improvement.[38]

Although these suggestions are subjective, and there is no stated definition of terms such as "supportive style": they should be seen as positive attributes for any management model. Optimally, these attitudes can be expressed in the Quality Policy.

Design

Skipping the sections of ISO 9004 that discuss financial considerations (paragraph 6) and marketing (paragraph 7), the next section of interest discusses quality in design. Obviously, this is a critical section for designers and engineers. As with the other sections discussed to this point, the information in ISO 9004 is not considered to be a requirement emanating from a standard that must be complied with to receive ISO 9001 registration. Instead, ISO 9004 constitutes a guideline or introduction to the concepts presented in the standards like ISO 9001 to ISO 9003. Since ISO 9001 is the only standard that covers design and development, clearly the information in ISO 9001 takes precedence over the presentation here. This presentation is intended to introduce the reader to the subject of the generic ISO 9000 quality systems and their relationship to the design process.

First, it should be clear that the design process is going to have an enormous impact on the achievement of quality. If quality is interpreted to mean consistently meeting customers' requirements, it is essential that the product be designed to be able to meet these requirements. ISO 9004 states this at the beginning of Section 8 which covers quality in specification and design. According to paragraph 8.1, the "design function should provide for the translation of customer needs into technical specifications for materials, products, and processes. This should result in a product that provides customer satisfaction at an acceptable price that gives a satisfactory financial return for the organization."[39]

Although different organizations approach design in different ways, each organization should be able to identify a design input which specifies the translation of the customer's requirements into the design function. Also, each organization that performs the design process is going to have a "design output" that is supposed to meet that input. The job of the quality system with respect to design is to see that the design output satisfies the design input (Figure 9.5). Furthermore, it is the job of the ISO 9001 standard to see that the quality system in fact does this.

In order to ensure that design output satisfies design input, ISO 9004 discusses some tools and guidelines for the management of the design process. Some of these tools include design planning, design review, design verification, design validation, and configuration management. These tools have a close relationship to ISO 9001. Although these are not the exact terms used in ISO 9004, they are the terms used in ISO 9001.

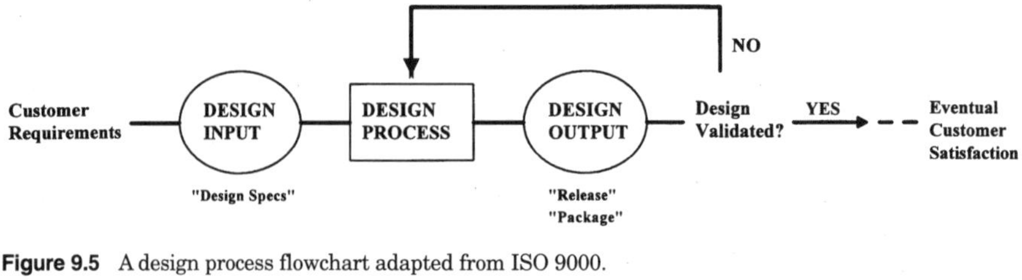

Figure 9.5 A design process flowchart adapted from ISO 9000.

Design planning

Design planning can be seen as the activities engaged in by engineering management prior to the start of a design process. Unfortunately, the distinction between the planning stage and the beginning of the design process is not always clear. In many cases, preliminary designs of various forms are discussed before a formal final design process commences. However, a number of advantages may be present if a clear distinction is made between these phases. First and foremost, the design planning stage can assist one in clearly specifying the design input. Since design input involves turning the customer's needs into actual specifications, the customer's involvement is critical. The sooner the needs of the customer are understood, the fewer revisions may be needed in later phases of work. This should lead to better productivity, higher financial return, and a faster time to market. Therefore, it should be advantageous to have a clearly specified period of design planning time prior to the beginning of a development project.

Another advantage of using a design planning scheme is that it facilitates preventative action with respect to the product and the design process. The design planning stage can be used to anticipate problems in the product specification. The design plan personnel can check whether the customer has thought about all aspects of their product proposal (safety, adherence to regulations, etc.). Customers may not always be as knowledgeable about these aspects as the design personnel within the design and engineering organization.

The design planning stage can also be used to anticipate problems in the product itself. The design plan personnel can check if production facilities are going to be incapable of meeting the customer's request (again based on the specialized knowledge possessed by the design and engineering personnel). The design plan personnel could also check if needed design tools (such as computers and software) are available to meet the demands of the customer's request. Regardless of the scenario, the design process is enhanced when design personnel spend some time in the design planning phase considering the task to be completed before work is started.

The design planning should establish responsibility for the design process. According to paragraph 8.2.1: "Management should prepare plans that define the responsibility for each design and development activity inside and/or outside the organization, and ensure that all those who contribute to design are aware of their responsibilities in relation to the full scope of the project."[40]

As the quality system defines the structure and tools of the organization for the achievement of quality, so the design organization must define the personnel and tools necessary for completing a particular design project. Responsibility should be assigned for each phase of the design and development process, including engineering, design, testing, and validation activities.

The next aspect of design planning is similar to project planning. The design plan should determine a schedule with milestones, holdpoints, etc. At these points, an assessment should be planned. According to paragraph 8.2.3: "Management should establish time-phased design programs with holdpoints appropriate to the nature of the product and process. The extent of each phase, and the position of the holdpoints at which evaluations of the product or the process will take place, can depend upon several elements."[41]

The design plan can help determine if adequate resources exist to carry out the design plan. By considering the design resources available and the proposed schedule, management should be able to determine if additional personnel or other resources are required.

Obviously, the best time to find out if there is a resource problem is before designing and engineering begins. In fact, at the design planning stage, management may easily get back to the customer with a revised schedule, whereas after the work is started it may be very difficult to revise a schedule. In terms of the holdpoints within the schedule, ISO 9004 suggests that factors such as "design complexity,...extent of innovation and technology being introduced,...the degree of standardization and similarity with past proven designs"[42] be used to assist in determining how often the holdpoints should be used.

The final guideline for the design planning is the use of acceptance criteria. A product may have measurable characteristics. If these characteristics are supposed to be within certain limits, the design planning process should finalize these limits. Specification of these limits is essential in guiding the design process. In turn, how successful the design output is in satisfying the design input can be judged based on these acceptance criteria. Again, it should be obvious that the best time to resolve these criteria is before the actual design process begins. Of course, in cases of strong process or product innovation (i.e., designing a product that has never been designed before), there may be little or no experience on which to base the development of acceptance criteria. However, this is not likely to be the most common scenario for design and engineering organizations. When this situation does arise, particular care must be given the entire design planning process.

In addition to specifying the acceptance criteria, it may be possible to specify the methods of measurement and test in the design plan phase. If not, then these should be specified during the design phase.

In some design projects, special hardware or software may be needed to measure the characteristics which must meet the acceptance criteria. Since this must be developed and obtained, it is best to consider this situation as well in the design plan phase. As stated in paragraph 8.3:

> The methods of measurement and test, and the acceptance criteria applied to evaluate the product and processes during both the design and production phases, should be specified. These should include the following:
>
> a) performance target values, tolerances, and attribute features;
> b) acceptance criteria;
> c) test and measurements methods, equipment, and computer software.[43]

Design review

The next tool described by ISO 9004 with respect to the design and development process is the design review. This is a critical issue since it is, in fact, a requirement in ISO 9001 for design verification. Basically, a design review is a meeting. However, in keeping with the ISO 9000 philosophy, this meeting must be part of the quality system procedures. As such, procedures would have to document at what point in a project the meeting is to be held, and it would have to document what parts of the organization attend the meeting. In addition to being called for in the procedures, this meeting must have a quality record to show that it has occurred. Most likely minutes of the meeting would be created, signed, and maintained in a filing system.

The design review is not a progress type of meeting. According to paragraph 8.4.1, "a formal, documented, systematic, and critical review of the design results should be planned and conducted. This should be distinguished from a project progress meeting."[44] A key description of the design review is that it is a "critical review." This meeting is meant to critique the design. Although there is a constant process of give and take and critique occurring in any designer's mind, there is some value to bringing various experts together for a formal meeting. Although it seems to contradict the trend for concurrent engineering (where these parties work together all the time anyway in the design process), there is no reason that concurrent engineering organizations can not hold a design review meeting. Hopefully, the design review meeting can at least provide an end to the design phase and reinforce the need for coordination between various facets of the organization which is going to be affected by the outcome of the design.

In terms of timing, ISO 9004 states that a design review should be held "at the conclusion of each phase of design development."[45] But ISO 9004 does not define any phases of design development (recall that it is not a standard, but a guideline). The timing clearly must be determined based on the type of product or project. Some possible phases would be preliminary analysis, prototype design, preliminary design, or final design.

Besides reviewing the design and identifying problem areas, the design review should be used to anticipate problems. As previously mentioned, it is clearly advantageous to anticipate problems early in the design development process. Thus, the participants in the meeting should be prepared to think about problems that may arise in the future (in terms of production, testing, safety, etc.).

ISO 9004 lists a number of items as elements of design reviews. These items are categorized as "items pertaining to customer needs and satisfaction," "items pertaining to product specification," and "items pertaining to process specification."[46] The first category is for critiquing the designs ability to meet the customer's needs. This part of the design review meeting should concentrate on comparing the design input (the customer specifications and perceptions) with the state of the design. This part of the design review is essential, and it must be documented.

Those attending the meeting should agree that the design is sufficiently likely to meet the customer's specifications, and that the next phase of design, development, or production can begin. This decision can be based upon the experience of those attending the meeting, but it should also consider use and/or success of prototype testing, comparison with previously successful and/or competitive designs, and compliance with industry or government guidelines or standards. According to paragraph 8.4.2a of ISO 9004, the following are elements of a design review (items pertaining to customer satisfaction):

1) comparison of customer needs...with technical specifications for materials, products, and processes;
2) validation of the design through prototype tests;
3) ability to perform under expected conditions of use and environment;
4) unintended uses and misuses;
5) safety and environmental compatibility;
6) compliance with regulatory requirements...;
7) comparisons with competitive designs;
8) comparisons with similar designs, especially analysis of the history of ...problems.[47]

Beyond these basic considerations for whether the product is going to satisfy the design input, the design review meeting should consider extraordinary conditions. Those attending the meeting should agree that the design is appropriately safe under normal, as well as unintended, and even misused conditions. ISO 9004 suggests considering failure mode and effect analysis, fault tree analysis, labels, warnings, user instructions, etc. Those attending the meeting should also agree that the design is appropriately environmentally compatible. ISO 9004 suggests considering storage needs, shelf life, fail-safe characteristics, labels, traceability, etc. According to paragraph 8.4.2b, the following are elements of a design review (items pertaining to specifications):

1) dependability and serviceability requirements;
2) permissible tolerances and comparison with process capabilities;
3) product acceptance criteria;
4) installability, ease of assembly, storage needs, shelf-life, and disposability;
5) benign failure and fail-safe characteristics;
6) aesthetic specifications and acceptance criteria;
7) failure mode and effect analysis, and fault tree analysis;
8) ability to diagnose and correct problems;
9) labeling, warnings, identification, traceability requirements, and user instructions;
10) review and use of standard parts.[48]

In addition to the quality of the design with respect to meeting external customer needs, the design review meeting can address a second category of items. These items pertain to the quality of the design with respect to meeting internal customer needs. Those attending the meeting should agree that the design has proper specifications and drawings for use by the rest of the organization. These specifications should have permissible tolerances, and they should be achievable by the resources (tools, people, etc.) scheduled to create the product (process capability). In conjunction with the product specification, those attending should agree that appropriate acceptance criteria have been specified.

The final category listed in ISO 9004 for items pertaining to design reviews is process specification. Depending on the organization, the personnel creating the design may or may not specify the processes to be used to create the product. If the processes are known at the time of the design review, then those attending the design review should agree on what if any special processes need to be developed for product assembly, automation, installation, etc.[49] In addition to production, any special needs for product inspection and test, specific materials, suppliers, subcontractors should be considered. In terms of safety and environmental impact, the design review meeting should consider these items with respect to internal procedures and equipment, shelf-life, disposal, etc.

As shown by ISO 9004, the content of design review meetings are going to have to be determined by the product and organization involved. In ISO 9001, the design review is a requirement, so a controlled procedure is going to be adopted which more specifically defines these meetings. This department procedure or quality system procedure should specify what parts of the organization should be represented, what internal and external standards should be checked for compliance, where the minutes of the meeting will be stored and controlled, and whatever else is appropriate to the product and organization.

Verification versus validation

The next two paragraphs of ISO 9004 cover what are basically the most important concepts for designers and engineers in the ISO 9000 area—design verification and design validation. Both these concepts are found in ISO 9001, and they are very important requirements for meeting ISO 9001. These concepts are often interchanged and they are often confused. The meanings of these terms presented here are based solely on an interpretation of ISO 9004 and ISO 9001. The meanings of these terms may be different in other fields of study or in different organizations. The author's background is in engineering software development and verification and validation do not, in fact, seem to be clearly defined in the field of software quality assurance.

In this book, design verification is used to ensure that the design fulfills the product specifications. In terms of ISO 9001, this can be interpreted as ensuring that the design output meets the design input. Design verification is conducted in the context of the design within the scope of design specifications. Designers and engineers should view design verification as a check on their work within their own environment. For design verification, the product should still be at the prototype stage.

This situation can be contrasted with the concept of design validation. Design validation is used to ensure that the design fulfills customer needs. Design validation is conducted in the context of the design within the scope of customer specifications. In many cases, this is a more macroscopic approach. Also, note that the product is at the final design or ready-for-production stage. Not only the work of designers and engineers is judged at this point, but also the work of those that translated the customer needs into design specifications. At the design validation phase, the design output may meet design input, but the design input may not be adequate for some reason.

Also, the results of the design validation process form an important body of documentation and information; it provides a benchmark for future revisions to the product.

This presented distinction between design verification and validation should be clear in looking at ISO 9001, which states that "design verification shall be performed to ensure that the design-stage output meets the design-stage input requirements,"[50] whereas "design validation shall be performed to ensure that product conforms to defined user needs and/or requirements."[51] Furthermore, it is stated that "Design validation follows successful design verification," "Validation is normally performed under defined operating conditions," and "Validation is normally performed on final product."[52]

Design verification

Turning to a more detailed discussion of the concept of design verification, keep in mind that design verification should be viewed as a check on the work of designers and engineers. Also, recall that the design review meetings mentioned earlier cover the same topic. Therefore, the design review meeting is a form of design verification. Indeed, ISO 9001 in the 1994 edition (unlike the previous 1987 edition) indicates a design review as part of design verification. The documentation from this meeting becomes an essential quality record.

More forms of design verification besides the design review meetings should be used as well. ISO 9004 states the following for design verification:

> design verification should include one or more of the following methods:
>
> a) performing alternative calculations...,
> b) testing and demonstrations (e.g., by model or prototype tests)...,
> c) independent verification, to verify the correctness of the original calculations.[53]

The demands for design verification are no more stringent in ISO 9001 which states the following:

> In addition to conducting design reviews..., design verification may include activities such as
>
> -performing alternative calculations,
>
> -comparing the new design with a similar proven design, if available, and
>
> -undertaking tests and demonstrations...[54]

Although there may be other methods of design verification to be considered, the ones mentioned so far should be relevant and useful in most design and engineering organizations.

The first method is alternative calculations. In most engineering analyses, there are a number of methods published or available for calculating specific parameters. Certainly in the area of stress analysis, there are a variety of accepted techniques for various situations such as normal force per area, beam theory, formulas for stress and strain, finite element analysis, boundary element analysis, etc. In these cases, more than one method can be used to calculate various types of stresses for various geometric locations induced by various anticipated forces. In terms of design verification, if one of these methods were used in the design process, one of the other methods could be used as a check on the original calculations. Naturally, the use of alternative calculations is going to be extremely dependent on the field of study or engineering involved, the product being designed, and the interpretation of the critical items to be verified. However, a great deal of insight into the design and its methods might be generated by satisfying the need for design verification with alternative calculations.

There are a number of factors to consider for alternative calculations. First, for design verification purposes, these alternative calculations must be documented. Hopefully all engineering calculations for a product are documented using a format and style dictated by department procedures or perhaps by the Quality Manual. These calculation sheets or computer records become the quality record of the basic calculations to be verified. The alternative calculations can then be performed in a similar manner. At this time, ISO 9001 can be interpreted as allowing the alternative calculations to be performed by the same designer or engineer that did the original calculations. However, some design and engineering organizations may wish to make use of a policy that indicates that alternative calculations should not be performed by the same person that performed the original calculations.

It should come as no surprise that many design and engineering calculations are now performed using computer software. However, an ISO 9001 audit can also require that this software be verified (depending on how critical the application is to the safety or quality of the product). In terms of verifying this software (particularly in-house software), alternative calculations are quite valuable. Just as alternative calculations of different types can be used to verify manual calculations, alternative calculations can be used to verify automated calculations. Indeed, in computationally intensive procedures (such as finite element analysis), which can hardly be duplicated manually, the use of a manual calculation using a simpler method can be a useful verification. In this case, a simple problem with a direct solution can be modeled using the software and then compared with the solution than has been derived for this simple or ideal case. Of course, to be useful for design verification, this software verification process and results need to be documented and controlled. This quality record can then be retrieved in the future as an engineering reference or as evidence during an audit.

The next method of design verification listed in ISO 9004 (as well as ISO 9001) is testing. According to ISO 9004, this is "testing and demonstrations (e.g., by model or prototype tests)."[55] As before, the intent is to check the design and engineering calculations. The testing at this stage is done with prototypes as opposed to production samples. In this stage, testing may make use of models in a laboratory environment as well. If models, environmental simulations, etc. are going to be used, this should be stated and reviewed in the design planning stage. During that planning there should be a documented rationale for permitting the use of the models or simulations. The limitations of the tests using these tools should also be clearly stated.

The results of the design verification tests can be compared to the calculations used in the design phase. Of course, the end result of the testing should be documentation. As paragraph 8.4.3 indicates, "if this method is adopted, the test programs should be clearly defined and the results documented."[56] These design verification test reports also form a valuable body of evidence for the internal and external audits.

The final method of design verification listed in ISO 9004 is called independent verification. According to ISO 9004, the purpose here is "to verify the correctness of the original calculations and/or other design activities."[57] In order to check the calculations of designers or engineers, the same calculation procedure (as opposed to alternative calculations) could be performed by an independent organization. This does not appear to be as useful as alternative calculations since the original method may be invalid (the wrong choice of design method may have been made). The independent calculation would only help show that the calculation method had been properly executed. Of course, this method of verification is only suggested and could be used in addition to testing and alternative calculations. As before, recall that the intent is to check the assumptions and calculations of designers and engineers. The result must be some form of documentation that can be stored and controlled for future reference.

Design validation

As mentioned earlier, design validation follows successful design verification. This does not mean that there should only be one validation, however. ISO 9004 states that the validation can occur at significant stages of the design. Design validation would be done in the context of user needs as opposed to design specifications. The intent is not to verify design and engineering calculations, but to see if the product designed meets the customer's needs.

ISO 9004 states that design "evaluation can take the form of analytical methods, such as FMEA (failure mode and effect analysis), fault tree analysis, or risk assessment, as well as inspection and test."[58] If testing is chosen, then ISO 9004 states that the "amount and degree of testing (see 8.3) should be related to the identified risks," "Independent evaluation can be used...," and that a "number of samples should be examined by tests and/or inspection to provide adequate statistical confidence in the results."[59] Furthermore, the testing should be in the user's environment. Paragraph 8.5 continues as follows:

tests should include the following activities:

a) evaluation of performance, durability, safety, reliability, and maintainability under expected storage and operational conditions;

b) inspections to verify that all design features conform to defined user needs....[60]

Of course, the results of any design validation testing should be documented.

Interestingly, the next sentence of paragraph 8.5 states that tests should include "c) validation of computer systems and software."[61] There are several possible meanings to this statement. ISO 9004 could be indicating that if computers are part of a physical product (such as electronic control systems), then the software should be validated as well as the physical product. A second possibility is that ISO 9004 could be indicating that if computer systems are used in the testing activity to gather and reduce measurement data (data acquisition), then these computer systems should be validated to be certain that collected data are valid. Finally, ISO 9004 could be indicating that if computer software is used in design and engineering calculations (as discussed under design verification), then this software should be validated.

The best interpretation of the statement under consideration is that all three scenarios are valid applications of ISO 9004. Indeed, with respect to the third scenario, the author recommends that in-house as well as commercial engineering software be validated after being loaded on user's computers. In this case, the organization's engineering software is reexecuted and its results compared to some benchmark results. Test data acquisition software (the second scenario) could also require periodic reevaluation against calibration data.

If testing is not chosen for design validation, qualified evaluators can apply analytical methods. As with the testing, the evaluation should be in the context of the user's environment and needs. And as always, the results "of all tests and evaluations should be documented regularly throughout the qualification test-cycle."[62]

Final design review

Paragraph 8.6, the next paragraph of ISO 9004, discusses a final design review. Although not discussed in ISO 9001, having a final design review meeting may be desirable, and it is included in the ideal ISO 9001 development process presented in Chapter 6 as the design review in the Design Output Stage. As already discussed, a design review meeting can be used as a form of design verification. The purpose of this meeting is to check the work of designers and engineers and to verify that the design specifications are being satisfied. In ISO 9001, design reviews are called for at "appropriate stages of design"[63] so there could be several meetings of this type during the design process. A final design review meeting can then be used at the end of the design phase (which encompasses the design input, design, and design output activities).

The final design review should be the milestone at which the design baseline is established. Subsequently, configuration management is used to track evolution of the design and subsequent phases in its life cycle. Furthermore, the completion of the final design review leads to an approval cycle. According to paragraph 8.6, the "total document package that defines the design baseline (output) should require approval at appropriate levels of management...."[64] If the approval cycle is completed successfully, ISO 9004 states that "this approval constitutes the production release and signifies that the design can be realized."[65] As with any policy concerning the quality system in the design and engineering organization, the procedure to follow for this final design review is to be covered by a documented procedure (either in the Quality Manual or a department procedure). Of course, the procedure for usual design reviews can also be used for the final design review as well (probably with the stipulation that the procedure specifies whether standard or final design review is being documented).

Market readiness review

Although market readiness review is not discussed in ISO 9001, it does provide another opportunity to assess the organization's ability to satisfy customer requirements prior to release of the product. Many of these aspects should already be addressed during the quality planning stage, but the organization's capabilities may have changed during the time elapsed for the design phase. In any case, ISO 9004 paragraph 8.7 suggests that a market readiness review meeting be held (and obviously documented). Points suggested to be considered includes status of manuals (assuming manuals are provided with the product), distribution channels, aftersales service, training of field personnel, field trials, tests, inspection of early production units, and process capability (i.e., checking if production equipment/personnel meet product specifications).[66]

Considering the points covered in the market readiness review, the personnel involved should go beyond the scope of the design and engineering staff. Even if a concurrent engineering concept is used to have design and manufacturing involved during the design stage, care must be taken to have sales and field support involved in the market readiness review. As with any policy concerning the quality system in the design and engineering organization, the procedure to follow for this review is to be covered by a documented procedure (either in the Quality Manual or a department procedure).

With the market readiness review coming after the design phase is complete, design changes may be needed after this review. One should note that this is a less than ideal time to find out that the design exceeds process capabilities, for instance, but this may happen. Assuming that the baseline for design output has already passed (at the final design review meeting), one may consider changes at this point to be a controlled revision of the design (as in an Engineering Change Release procedure). Therefore, as mentioned earlier, the design phase is considered complete and the market readiness review is really a process of a postdesign phase.

Design change control

The paragraphs remaining in the design section of ISO 9004 deal with more processes that come after the design phase is complete for a particular revision of a particular product. One such topic of this type is design chage or revision control.

For many reasons, a given completed design may require modification. As already mentioned, the market readiness review may trigger a design revision. As production processes, personnel, customer specifications, government regulations, or technologies change there may also arise a need for a design revision. Although it is best to avoid revisions, they are going to be unavoidable in a large organization with controlled work procedures. With controlled work procedures, there will be some boundary established between groups in the organization and between the organization and the customer. As the design progresses beyond the design and engineering group, new requirements are likely to be discovered. Also, new uses of the design may be discovered, and a design revision can allow the organization to take advantage of new uses of the design.

Regardless of the cause, it is essential that changes to the design be made in a controlled, documented manner. According to ISO 9004 paragraph 8.8, the "quality system should include documented procedures for controlling the release, change, and use of documents that define the design input and design baseline (output).[67] This statement should come as no surprise since all procedures affecting the quality of the product are supposed to be documented. Procedures such as the design revision should probably be part of the Quality Manual since it will probably be accessed by various groups in the organization (beyond the scope of design and engineering).

The design revision process is also likely to be centered on drawings. In this context, drawings means all forms of documentation that is necessary to manufacture a product. This definition can include traditional drawings showing views of parts, components, assemblies, arrangements, etc. Drawings could also include bills of materials, assembly instructions, tables of data, schematics, etc. Although ISO 9000 is intended to apply to all kinds of activities, this section of ISO 9004 is clearly oriented toward manufacturing (fortunately for designers and engineers). Thus, when the design was first considered complete (after the final design review), the drawings should have been given an approval, considered at a baseline revision level, and fixed in the configuration management scheme. For simplicity, these drawings can be considered at revision 0, or "rev level 0."

As stated earlier in the quote from paragraph 8.8, the design revision procedure needs to address the changes to these drawings. Who is to request changes? How are such changes to be documented? One possible solution is to use a hard copy or electronic form such as an Engineering Change Request. In this case, the appropriate personnel (specified by the controlled, quality manual procedure), would indicate to the design and engineering group that a change is needed on a specified drawing or document. This form would then be sent to the appropriate personnel who must then process the change request. The details of this process is obviously dependent on the nature of the organization.

After the design changes are made, the corrections are to be approved. As found in paragraph 8.8, "The procedures should provide for various necessary approvals."[68] This is a common requirement in the use of drawings. As the drawings were originally produced, one can assume that there was an approval requirement where the drawings are signed by the appropriate supervisor, engineer, etc. At each new revision level this approval procedure is repeated.

When any changes to drawings have been completed, the overall design revision procedure is not yet finished. The organization's design revision procedure must allow for removing the superseded drawings. Furthermore, internal quality auditors should be checking that obsolete drawings are not being used. As paragraph 8.8 continues regarding design-change control, "the procedures should provide for various necessary approvals, specified points and times for implementing changes, removing obsolete drawings and specifications from work areas, and verification that changes are made at the appointed times and places."[69]

Another problem to be addressed by the design revision procedure is whether or not the design needs to be revalidated. Although the design revision procedure can apply to the need to correct or upgrade drawings, this same procedure may also be used to fundamentally change the nature or performance of the design itself. In this case, the design validation must be repeated. According to paragraph 8.8, "Consideration should be given to instituting formal design reviews and validation testing when the magnitude, complexity, or risk associated with the change warrant such actions."[70] Obviously each organization is going to have its own criteria for triggering a design review, but a consistent documented procedure should still be developed.

In terms of revalidation, recall that the completion of the design stage resulted in a baseline for the design with documentation for the product's performance. This is essential in view of the design review triggered by the design revision procedure. This is due to the fact that although a particular design change seems reasonable to correct a localized problem, there may be unforeseen global problems caused by the design change. The baseline performance results can now be used to check that the product still performs acceptably after the change. This is a requirement that may be new to organizations that have not had prior experience with quality system standards such as ISO 9001. As shall be seen in paragraph 8.9 of ISO 9004, these baseline results are going to be needed even if there never are design revisions for a product.

Yet another problem to be addressed by the design revision procedure, is the need for "emergency" changes. According to paragraph 8.8, "these procedures should handle emergency changes necessary to prevent production or delivery of nonconforming product."[71] In the case where changes need to be made directly, an organization can have production of a product altered prior to the revision and release of new drawings. However, an approval procedure should still remain and be documented. In the case of the emergency revision, only the revision and issuance of the new drawings and removal from work areas is postponed. The same check procedures should apply, and if the design revalidation procedure is triggered, the design changes should be postponed (unless the design can be revalidated on a more immediate basis).

Design requalification

As stated in the earlier discussion on design change control, a design review (such as that used during and at the completion of the design stage) may be needed after a design revision. The purpose of this review is to check that the design is still properly meeting customer needs in view of changes in the design. Although not mentioned in the ISO 9001 standard, paragraph 8.9 of ISO 9004 goes beyond this requirement. In this paragraph it is stated that "periodic evaluation of product should be performed in order to ensure that the design is still valid."[72] This is termed design requalification.

Design requalification is performed at regular intervals. It is done whether there are design revisions or not (although the design review procedure used may be identical to the procedure used for design revisions). Paragraph 8.9 mentions some nondesign issues which should be addressed in this requalification. This includes " review of customer needs," "technical specifications in light of field experiences," and "new technology and techniques."[73] This periodic review process is performed to ensure that the design is still valid considering these external parameters (as well as internal process improvement factors).

An important facet of requalification is finding unexpected affects on a product's quality. In other words, as the design evolves through revisions, one can consider whether original performance requirements have been relaxed. If this is so, does the new performance still meet the customer's needs? Continuing in paragraph 8.9, "care should be taken that design changes do not cause degradation of product quality for example, and that proposed changes are evaluated for their impact on all product characteristics in the original product specification."[74] Obviously the baseline results documentation produced at the completion of the design stage (as mentioned earlier) is essential in making this judgement. Although the baseline results may be the original baseline after the design stage, the baseline results may also be new baseline results created after a design review triggered by a substantial design revision procedure. The control of these baseline results should be addressed by the configuration management.

ISO 9004 does not indicate how often this requalification should occur. In the author's experience with engineering computer programs, a reevaluation every 6 months was deemed acceptable by an ISO 9001 audit team. In other words, every 6 months all the engineering organization's analytical computer programs had to be rerun and compared to benchmark results. Obviously, the length of the product life cycle should be considered. If an organization's products are developed, revised, or decommissioned within a year time frame, then the requalification could occur monthly. If an organization's products are developed, revised, etc. within a time frame of decades, then probably yearly requalification could be used.

Configuration management

The next paragraph of ISO 9004 (paragraph 8.10) is for configuration management in design. Paragraph 5.2.6 of ISO 9004 discussed configuration management as well. In terms of configuration management in design, paragraph 8.10 merely indicates that it can be "most useful during the design phase." And, that it "continues through the whole life-cycle of a product."[75]

Configuration management is the tool by which an organization tracks the life cycle of a product. Configuration management should allow the organization personnel to know how many products are currently available and what phase of the life cycle each of those products occupies (design, production, installation), etc. Configuration management in design may take several forms, but is generally concerned with master or index documentation. Although many drawings may be needed to produce a product, each of them should have a revision level. Configuration management for these drawings could take the form of a hard copy or electronic index of all drawings needed for a product and the revision level. This is a master document which must be controlled and maintained in conjunction with the phase of the product (whether in the design, production, or installation and servicing phase).

As mentioned in the preceding discussion on design revision and design requalification, it is critical to maintain and control records on the status of each product and its drawings (drawings again implying other product design documentation as well). The status of the documentation allows the organization to know what the current disposition of a design. This is obviously essential to checking that design revisions and requalifications are valid.

Of course, design configuration management is also necessary to ensure that the proper revision level of a design is being produced. It should be constantly checked my means of internal audits that the proper revision level of drawings is being used in the production phase. External auditors are certain to check the production area for the disposition of drawings. These drawings must be in agreement with the configuration management records.

Inspection and test

Paragraph 13.1 of ISO 9004 is of interest to both designers and engineers. This paragraph is found in the section on control of inspection, measuring, and test equipment. Obviously, a variety of instruments and techniques may be used in laboratory or field environments during the design and development process. Products may be tested for design verification, design validation, design revalidation, etc. In each of these cases, it is important to trust the results that are being obtained.

According to paragraph 13.1, "control should be maintained over all measuring systems used in the development, production, installation, and servicing of product to provide confidence in decisions or actions based on measurement data."[76] The use of proper calibration and control procedures is essential in attaining this confidence. The need for this control should be clear. In the case of design verification, laboratory tests may be used to check the results that have been attained by designers or engineers. Therefore, the accuracy of the test results are directly related to the validity of the design verification.

Some of the items mentioned by ISO 9004 that should be controlled are "gauges, instruments, sensors, special test equipment, and test software."[77] All laboratory or field test equipment used by an organization should already be in an controlled environment of some sort. If not, this control must be established prior to any sort of audit such as ISO 9001 that is to cover the product development process. The control should include calibration to ensure that the readings of the equipment are in agreement with traceable sources. It should include procedures to ensure that the calibrated equipment remains in the calibrated state. Also, it should include a proper physical environment to ensure that equipment remains in the calibrated state and that unauthorized use or misuse is prevented.

Note that this paragraph 13.1 of ISO 9004 also mentions test software. In the development process, this test software should include data acquisition systems. These computer systems translate sensor or transducer readings into raw and reduced data for analysis. The test software should be "calibrated" or validated as well. This process should be documented to ensure that the electronic form of the readings agree with observed values. As mentioned previously concerning engineering analysis software, the data acquisition software should also be revalidated at regular intervals.

As with all the elements of the quality system, there should be documented procedures covering the processes used in the development data acquisition environment. According to paragraph 13.1, these documented procedures "should be established to monitor and maintain the measurement process."[78] Later in the paragraph, it is stated that the use of test equipment is to be in "conjunction with documented procedures to ensure that measurement uncertainty is known."[79]

The operating procedures of the laboratory or test environment would likely take the form of department procedures (i.e., middle-level documentation). These procedures would outline how equipment is to be used, what calibration procedures are to be used, safety precautions, software instructions, etc. The calibration procedure must also create documentation that shows what equipment has been calibrated, by whom, and at what time. This documentation would be considered a quality record (or bottom-level documentation), and it would be needed for audit situations.

There should also be a procedure for preparing reports that document the results gathered. These reports or data reduction form another quality record. These results can then be used in a design verification or design validation or design requalification review.

In the previous discussion, the use of statistical technique is avoided. However, paragraph 13.1 states that documented procedures should be used to keep the measurement process in a state of "statistical control, including equipment, procedures, and operator skills."[80]

This item has been avoided since the analogous paragraph in ISO 9001 (4.11) does not mention the need to use statistical methods directly. Instead, paragraph 4.20.1 of ISO 9001 states that the supplier (or product design organization in this case) "shall identify the need for statistical techniques required for establishing, controlling, and verifying process capability and product characteristics."[81] It is not clear, then, if statistical techniques are a requirement for testing in ISO 9001. Regardless of whether it is required or not, it would be good practice to compile the results of calibration of equipment and the effectiveness of personnel performing test activities. Monitoring the amount of recalibration required to keep an instrument within specifications might be valuable in predicting when an instrument is going to fail or need maintenance, for instance.

Corrective Action

Another important component of any quality system that is to meet the ISO 9000 requirements is corrective action. As one might expect, there are going to be nonconformances in the execution of a quality system. A nonconformance could be failing to meet the requirements of the quality system internally or failing to meet the requirements of the customer. In any case, it is essential that an organizational procedure (most likely as part of the Quality Manual) be created to address these activities.

It should be pointed out that many quality programs stress the importance of preventative action or striving for error-free work (as opposed to correcting problems after they are found). This is necessary since the members of an organization may decide to fall back on the corrective action system to handle problems in the future (as opposed to addressing them during early stages of a product life cycle). However, the striving for total quality is not likely to permit an organization to drop the corrective action procedure. Instead, one would hope that the higher level of quality would keep the resources applied for corrective action to a minimum.

According to ISO 9004, then, "corrective action begins with the detection of a quality-related problem and involves taking measures to eliminate the recurrence of the problem."[82] Indeed, the occurrence of a problem is an excellent time to examine all aspects of a process or product. This is a time to use quality-related tools such as "root cause," "cause and effect," "fish bone" (a technique for examining and graphing sources of problems), etc. or other quality management techniques. As mentioned previously, there should be an attempt to solve the existing problem while probing for other areas that may need improvement. This aspect of the corrective action is preventative in nature.

With respect to corrective action, ISO 9004 specifies that there should be assignment of responsibility and authority. This assignment is similar to the assignment of overall responsibility for the quality system (as seen in paragraph 5.2.2 of ISO 9004). In the case of paragraph 15.2, it is clear that the responsibility of the quality function in the organization includes corrective action. This is consistent with ISO 9001 as well. Therefore, an organization must be certain to include corrective action as a requirement of the quality function of the management organization in preparation for an ISO 9001 audit.

Although the quality function of the organization may be responsible for coordinating and managing the corrective action, many other parts of the organization are going to be affected by the corrective action process itself. Thus, if a quality problem is found in the design or engineering area, the design and engineering department should implement the corrective action. However, it is the quality function that is to record the nonconformance, record the part of the organization assigned to correct the problem, record the progress of the action, and record the successful completion of permanent corrective action. All this information becomes the quality record that is supposed to demonstrate that a corrective action system is in place and functioning.

Quality Records

As mentioned earlier, quality records are the evidence or output from the quality system procedures. An entire section (paragraphs 17.1–17.3) of ISO 9004 is devoted to quality records. Organizations generate quality records of their own design according to their own quality system, but ISO 9004 mentions a number of types of records that would be controlled as quality records. Many of these items are of interest to engineers and designers. Specifically, this list includes inspection reports, test data, qualification reports, validation reports, and calibration data. These items are to be controlled for "identification, collection, indexing, access, filing, storage, maintenance, retrieval, and disposition."[83]

Paragraph 17.3 of ISO 9004 also lists items of documentation that require control. This includes items that are not quality records but are of particular interest to designers and engineers. This list includes drawings, specifications, inspection procedures, test procedures, quality manual, quality plans (and thus design plans), operational (or departmental) procedures, and quality system procedures. Of course, a given organization may have other types of documentation that require control depending on the quality system that is implemented (such documentation may be needed to meet other standards besides ISO 9001).

In addition to creating and filing these records, a retention policy should be established. That is, the quality system documentation should specify how long the quality records should be retained. According to paragraph 17.2, "records should be retained for a specified time, in such a manner as to be readily retrievable for analysis, in order to identify trends in quality measures and the need for, and the effectiveness of, corrective action."[84] In light of the corrective action requirement of ISO 9001, and the reference to statistical methods in ISO 9001, instituting a record retention policy is clearly a recommended course of action.

Although there is a cost associated with the keeping of records, they can provide a valuable source of statistical data. The quality records can only provide these data if they are properly controlled. The advent of various computer-based document storage and retrieval systems can be useful in this regard. According to ISO 9001, paragraph 4.16, records "may be in the form of any type of media, such as hard copy or electronic media."[85] Of course, while in storage "quality records should be protected in suitable facilities from damage, loss and deterioration (e.g., due to environmental conditions)."[86] In terms of electronic media, this means that computer system must have documented system administration and backup procedures in place. Once quality records are stored in a computer-based system and a proper database or index is created for the quality records, they should be able to show interesting statistical information over time on the performance of the design and engineering organization and even the entire quality system.

Product Safety

The final aspect of the ISO 9004 quality system guidelines to be presented in this chapter is product safety. Most designers and engineers should be involved with this aspect of the quality system. Some of the guidelines from ISO 9004 are the following:

   a) identifying relevant safety standards in order to make the formulation of product specifications more effective;

   b) carrying out design evaluation tests and prototype (or model) testing for safety...;

   c) analyzing instructions and warnings to the user, maintenance manuals, and labeling.[87]

Each of these guidelines has a relationship to the design process. Each guideline may also be addressed as part of the design verification activity. If a safety requirement is part of the design specification, and the design verification activity is to check that the output from the design process meets the design specification, the design verification can also verify that the safety requirements are met. One possible exception to this approach is the third item shown. If the preparation of user instructions, manuals, etc. is done by a technical publications department that is separate from the design and engineering department, analyzing this material before release to the end user would have to be covered by a review meeting outside the scope of the design process.

In terms of the first product safety guideline item shown, identifying relevant safety standards should be met by the design specification document. This document should include specific safety needs in addition to the regulatory requirements and the acceptance criteria (or safety requirements could be made acceptance criteria). Assuming this information is included in the design specification, a design review meeting (in the design verification activity or the final design review) can include a review of the safety requirements in the design specification with respect to the current or final configuration of the design.

The second guideline for product safety (safety tests) can be incorporated into design verification as well. If there is a safety requirement in the design specification and testing is done for design verification, safety tests could be conducted as well. Of course, these safety tests could be conducted even if prototype testing is not used in the design performance type of testing (i.e., special safety testing). One advantage of doing safety tests before the final design is that changes to the design for safety consideration should be made more easily.

For the final guideline (reviewing or analyzing user instructions), the design review meeting activity can easily be used to meet this guideline. Assuming that this material is created with the product design, the design reviews can simply review this material and discuss the safety implications related to that material. If the design is released to production before this material is finished, the material should be reviewed in a separate process.

Conclusion

This chapter has presented a selective overview of the ISO 9004 guideline for quality systems with emphasis on technical professionals such as designers and engineers. You should now have an appreciation of the overall processes involved with establishing and maintaining a quality system. This overview is intended to help clarify concepts that are going to be found in the ISO 9001 standard for activities such as design specifications, design reviews, design verification, design validation, testing, and product safety. Of course, these concepts are not guidelines in ISO 9001; in ISO 9001 these types of activities are requirements. Failure to demonstrate that these activities are being conducted should result in a failure to achieve ISO 9001 registration.

References

1. ASQC, *ISO 9004 (ANSI/ASQC Q9004-1-1994) Quality Management and Quality System Elements—Guidelines,* ASQC, Milwaukee, Wisconsin, 1994.
2. ASQC, *ISO 9004 (ANSI/ASQC Q9004-1-1994) Quality Management and Quality System Elements—Guidelines,* ASQC, Milwaukee, Wisconsin, 1994, paragraph 4.4.1, p. 3.
3. ASQC, op-cit, paragraph 4.1, p. 2.
4. ASQC, op-cit, paragraph 4.4.2, p. 3.
5. ASQC, op-cit, paragraph 4.3.1, p. 2.
6. ASQC, op-cit, paragraph 5.1.2, p. 3.
7. ASQC, op-cit, paragraph 5.2.1, p. 3.
8. ASQC, op-cit, paragraph 5.2.2, p. 3.
9. ASQC, op-cit, paragraph 5.2.2, p. 3.
10. ASQC, op-cit, paragraph 5.2.2, p. 3.
11. ASQC, op-cit, paragraph 5.2.2, p. 3.
12. ASQC, op-cit, paragraph 5.2.2, pp. 3–4.
13. ASQC, op-cit, paragraph 5.2.3, p. 4.
14. ASQC, op-cit, paragraph 5.2.3, p. 4.
15. ASQC, op-cit, paragraph 5.2.4, p. 4.
16. ASQC, op-cit, paragraph 5.2.4, p. 4.
17. ASQC, op-cit, paragraph 5.2.5, p. 4.
18. ASQC, op-cit, paragraph 5.2.5, p. 5.
19. ASQC, op-cit, paragraph 5.2.5, p. 5.
20. ASQC, op-cit, paragraph 5.3.1, p. 5.
21. ASQC, op-cit, paragraph 5.3.2.4, p. 5.
22. ASQC, op-cit, paragraph 5.3.2.4, p. 5.
23. ASQC, op-cit, paragraph 5.3.1, p. 5.
24. ASQC, op-cit, paragraph 5.3.2.3, p. 5.

25. ASQC, op-cit, paragraph 17.3, p. 19.
26. ASQC, op-cit, paragraph 5.3.3, p. 5.
27. ASQC, op-cit, paragraph 5.3.3, p. 5.
28. ASQC, op-cit, paragraph 5.3.3, p. 5.
29. ASQC, op-cit, paragraph 5.3.3, p. 5.
30. ASQC, op-cit, paragraph 5.3.3, p. 5.
31. ASQC, op-cit, paragraph 5.4.1, p. 6.
32. ASQC, op-cit, paragraph 5.4.1, p. 6.
33. ASQC, op-cit, paragraph 5.4.2, p. 6.
34. ASQC, op-cit, paragraph 5.4.3, p. 6.
35. ASQC, op-cit, paragraph 5.4.5, p. 6.
36. ASQC, op-cit, paragraph 5.5, p. 6.
37. ASQC, op-cit, paragraph 5.6, p. 7.
38. ASQC, op-cit, paragraph 5.6, p. 7.
39. ASQC, op-cit, paragraph 8.1, p. 8.
40. ASQC, op-cit, paragraph 8.2.1, p. 9.
41. ASQC, op-cit, paragraph 8.2.3, p. 9.
42. ASQC, op-cit, paragraph 8.2.3, p. 9.
43. ASQC, op-cit, paragraph 8.3, p. 9.
44. ASQC, op-cit, paragraph 8.4.1, p. 9.
45. ASQC, op-cit, paragraph 8.4.1, p. 9.
46. ASQC, op-cit, paragraph 8.4.2, pp. 9–10.
47. ASQC, op-cit, paragraph 8.4.2a, p. 9.
48. ASQC, op-cit, paragraph 8.4.2b, pp. 9–10.
49. ASQC, op-cit, paragraph 8.4.2c, p. 10.
50. ASQC, *ISO 9001 (ANSI/ASQC Q9001-1994) Quality Systems—Model for Quality Assurance in Design, Development, Production, Installation, and Servicing,* ASQC, Milwaukee, Wisconsin, 1994, paragraph 4.4.7, p. 4.
51. ASQC, ISO 9001, paragraph 4.4.8, p. 4.
52. ASQC, ISO 9001, paragraph 4.4.8, p. 4.
53. ASQC, ISO 9004, paragraph 8.4.3, p. 10.
54. ASQC, ISO 9001, paragraph 4.4.7, p. 4.
55. ASQC, ISO 9004, paragraph 8.4.3b, p. 10.
56. ASQC, ISO 9004, paragraph 8.4.3, p. 10.
57. ASQC, ISO 9004, paragraph 8.4.3c, p. 10.
58. ASQC, ISO 9004, paragraph 8.5, p. 10.
59. ASQC, ISO 9004, paragraph 8.5, p. 10.
60. ASQC, ISO 9004, paragraph 8.5, p. 10.
61. ASQC, ISO 9004, paragraph 8.5, p. 10.
62. ASQC, ISO 9004, paragraph 8.5, p. 10.
63. ASQC, ISO 9001, paragraph 4.4.7, p. 4.
64. ASQC, ISO 9004, paragraph 8.6, p. 10.
65. ASQC, ISO 9004, paragraph 8.6, p. 10.
66. ASQC, ISO 9004, paragraph 8.7, pp. 10–11.
67. ASQC, ISO 9004, paragraph 8.8, p. 11.
68. ASQC, ISO 9004, paragraph 8.8, p. 11.
69. ASQC, ISO 9004, paragraph 8.8, p. 11.
70. ASQC, ISO 9004, paragraph 8.8, p. 11.
71. ASQC, ISO 9004, paragraph 8.8, p. 11.
72. ASQC, ISO 9004, paragraph 8.9, p. 11.
73. ASQC, ISO 9004, paragraph 8.9, p. 11.
74. ASQC, ISO 9004, paragraph 8.9, p. 11.
75. ASQC, ISO 9004, paragraph 8.10, p. 11.
76. ASQC, ISO 9004, paragraph 13.1, p. 15.
77. ASQC, ISO 9004, paragraph 13.1, p. 15.
78. ASQC, ISO 9004, paragraph 13.1, p. 15.
79. ASQC, ISO 9004, paragraph 13.1, p. 15.
80. ASQC, ISO 9004, paragraph 13.1, p. 15.

81. ASQC, ISO 9001, paragraph 4.20.1, p. 9.
82. ASQC, ISO 9004, paragraph 15.1, p. 17.
83. ASQC, ISO 9004, paragraph 17.1, p. 18.
84. ASQC, ISO 9004, paragraph 17.2, p. 18.
85. ASQC, ISO 9001, paragraph 4.16, p. 9.
86. ASQC, ISO 9004, paragraph 17.2, p. 18.
87. ASQC, ISO 9004, paragraph 19, p. 20.

10

The ISO 9001
Quality System

Introduction

Building on the basic concepts about quality systems from the discussion of ISO 9004 presented by the previous chapter, this chapter presents detailed information on quality systems for the ISO 9001 standard. ISO 9001 is the critical quality standard for designers and engineers. ISO 9001 makes similar statements to ISO 9004, but requirements are given in ISO 9001 instead of just guidelines.

ISO 9004 does, however, introduce the reader to many concepts found in ISO 9001, and you are encouraged to review ISO 9004 if you do not understand basic quality system concepts when reading this chapter. Of course, in this discussion of ISO 9001, the emphasis is placed on the design, engineering, and product development aspects of this standard. Indeed, the reader should be aware that some sections of ISO 9001 are not discussed at all by this book because they are not directly related to the activities of designers and engineers; however, these other sections will still be vital to the overall ISO 9001 effort for an organization.

The current designation of the ISO 9001 standard in the United States is ANSI/ASQC Q9001-1994. However, this book refers to it generically as ISO 9001. This standard is titled *Quality Systems—Model for Quality Assurance in Design, Development, Production, Installation and Servicing*. The reader should be aware that there are other quality standards besides ISO 9001 that are part of the ISO 9000 series of standards. However, only ISO 9001 encompasses the processes of design and development; therefore, it is ISO 9001 which is of most interest to the majority of designers and engineers.

The first paragraph of ISO 9001 defines the scope of the standard. It states that the standard is "for use where a supplier's capability to design and supply conforming product needs to be demonstrated."[1] The supplier in this case is the organization that is going to be designing and producing a product. In order to produce this product in a manner conforming to contract requirements or satisfying customer requirements, the supplier is to follow the requirements of ISO 9001, the assumption being that if the organization meets the requirements of ISO 9001, then that organization can design and produce quality products.

ISO 9001, as indicated by its title, is concerned with specifying requirements for a quality system. As discussed previously, a quality system is composed of an organizational structure, documented procedures, and tools (refer to Chapters 8 and 9 for a more thorough exploration of quality systems). The goal of this chapter on ISO 9001, then, is to present attributes of the organization's structure, procedures, and/or tools that must be present in order to satisfy the requirements of ISO 9001. Indeed, virtually all of ISO 9001 is devoted to presenting specific quality system requirements.

Quality Policy

The starting point for the ISO 9001 quality system is the Quality Policy. Although ISO 9004 states that there should be a Quality Policy, ISO 9001 says that the organization shall establish a Quality Policy. Therefore, if an organization is to be registered to the ISO 9001 standard, there must be a Quality Policy.

According to paragraph 4.1.1 of ISO 9001, "management with executive responsibility shall define and document its. policy for quality, including objectives for quality and its commitment to quality."[2] This paragraph also states that the organization "shall ensure that this policy is understood, implemented, and maintained at all levels of the organization."[3] Although these statements imply a hierarchy structure to the organization, any organizational structure should include personnel with executive or legal authority of some kind for the organization. These managers are expected to create and to disseminate a written policy that specifies the quality objectives of the organization. This would include a commitment to comply with standards such as ISO 9001 and a commitment to correcting any nonconformances with standards or statutes in a timely manner.

The Quality Policy is to be included in the organization's Quality Manual. This manual is controlled to ensure that all members of the organization can refer to the latest Quality Policy. This procedure ensures that personnel which affect quality have access to the Quality

Policy and are aware of its requirements. Considering the importance of design to product quality, one should assume that all designers and engineers in an organization are considered personnel who affect quality. Therefore, all designers and engineers should be familiar with the location and disposition of the Quality Policy at all times. Finally, in order to signify approval of the Quality Policy, the top-level or executive management of the organization is to sign the Quality Policy document.

Management Model

As organizational structure is one of the components of a quality system, clearly ISO 9001 places requirements on the organizational structure. This structure in concert with controlled and documented procedures should constitute a management model. Although "management model" is not a term used by ISO 9001, such a term may be useful for engineering and design personnel viewing ISO 9001. Organizational structure and procedures may be viewed as closely integrated, and the term "management model" is assumed to encompass both concepts. This should simplify the overall study of the quality system and its management.

Structure

One of the first management model requirements is to have clearly defined organizational structure. This should clearly apply in the design and engineering portion of the organization. According to paragraph 4.1.2.1, "responsibility, authority, and the interrelation of personnel who manage, perform, and verify work affecting quality shall be defined and documented."[4] Although this does not specifically state that designers and engineers must have a clearly defined position within the organization, it should be understood that designers and engineers involved with a product's development should always have a direct impact on product quality (both in terms of producing viable designs and in terms of meeting external customer's requirements). It is hard to imagine consistently meeting the demands of customers without considering designers and engineers as thoroughly vital links in the process of a quality system.

If, in fact, the work of designers and engineers affects quality, their organization (department, group, etc.) must be structured and documented as an ISO 9001 requirement. Their organizational structure also includes a definition of their responsibilities and authorities. For ISO 9001, these definitions must include the freedom to initiate preventative action, to identify and record problems, initiate corrective action, verify correction of problems, and control handling of nonconforming (or defective) product until it is corrected or discarded. These activities are covered in paragraph 4.1.2.1 as follows:

> personnel who manage, perform, and verify work affecting quality shall be defined and documented, particularly for personnel who need the organizational freedom and authority to:
>
> a) initiate action to prevent the occurrence of any non-conformities relating to product, process, and quality system;
>
> b) identify and record any problems relating to the product, process, and quality system;
>
> c) initiate, recommend, or provide solutions through designated channels;
>
> d) verify the implementation of solutions;
>
> e) control further processing, delivery, or installation of nonconforming product until the deficiency or unsatisfactory condition has been corrected.[5]

These quality-affecting activities are likely to be consistent with the operating practices of any design and engineering organization. Although designers and engineers may not have the authority in all cases to initiate or prevent production and design activities, they should have the authority to bring problems to light, point out preventative action, and keep their own records and correspondences in this regard.

Of course, in light of the ISO 9001 requirements, organizations should carefully review the organizational freedom of designers and engineers to be certain that the ISO 9001 requirement is met. Section 4.1.2.1 should be read and carefully applied by individual organizations. Item c) states there must be freedom to "initiate, recommend, or provide solutions through designated channels."[5] However, "designated channels" may be interpreted differently by different organizations and different auditing agencies. This paragraph should also be able to encompass concurrent engineering or other cross-functional teams through a team or group leader.

Note that design and engineering organizations may already have documented procedures for giving members of the design and engineering organization proper authority. For instance, engineering procedures may already exist for identifying nonconforming product or material, providing guidance and liaison to shop personnel for incoming material, requesting engineering or design changes, and requesting changes to design procedures, computer programs, etc. All these typical engineering activities should indicate the presence of authority over quality issues within the organization.

Management representative

As part of the management model, ISO 9001 requires a management representative. This is a person who is assigned the basic responsibility for the quality system. According to paragraph 4.1.2.3, this person is to ensure that a quality system is established, implemented, and maintained. Also, according to this paragraph, this person is appointed by management with "executive responsibility."

The management representative could be the manager of the quality assurance or quality control department of the organization; it could also be a higher-level executive such as vice president of quality or vice president for Total Quality Management. In any case, irrespective of his or her other duties, this person must have the authority to manage the quality system.

Finally, this person is required by ISO 9001 to report on the status or performance of the quality system. This review with management is to be the "basis for improvement of the quality system."[6] According to paragraph 4.1.3, a formal review of the quality system must be held with "management with executive responsibility"[7] at specific intervals. This formal review is to be documented so that records are maintained to show that the review is being held. At this review, the management representative can present the number of problems or noncompliances detected during internal audits, the number of quality manual procedures revised, etc. during the previous time interval.

Resources

The next aspect of the management model required is "adequate resources." According to paragraph 4.1.2.2, resource requirements are to be identified and then provided. Resources, in this case, are assumed to mean the personnel and material necessary to implement the quality system. Therefore, for ISO 9001 compliance, it is necessary to not only have an acceptable quality system documented, but the resources needed to make that quality system function must also be provided. Paragraph 4.1.2.2 mentions the need for "assignment of trained personnel for management, performance of work, and verification activities including internal quality audits."[8] Thus, the management model is required to have a sufficiently large quality assurance or quality control function so that it is capable of performing these activities.

The reference to the quality assurance staff may not be a major concern for the design and engineering department of an organization. In the design and engineering department, management, process capability, level of performance of work, and verification activities are often carried out by the design and engineering department itself (as opposed to being provided by the quality assurance staff). Although the internal quality audits might be performed by members of the quality assurance department, the design and engineering staff may also conduct its own informal audits or have liaison functions with the quality assurance department.

Quality Manual

In the discussion of ISO 9004, the concept of a Quality Manual was presented. This is a so-called top-level documentation that contains the most fundamental policies and procedures for the organization. Although ISO 9004 says an organization should have a Quality Manual, ISO 9001 paragraph 4.2.1 states that the organization "shall prepare a quality manual covering the requirements of this American National Standard."[9] Thus, it is clear that such a document must exist.

The Quality Manual is the obvious place to include the Quality Policy, the organizational structure, and the management representative definition and responsibilities. Depending on the size of the organization, the Quality Manual may also include the actual operating procedures of groups or departments within the organization. In other cases, the Quality Manual makes reference to lower-level documents to find these procedures. However, as stated in paragraph 4.2.2, these quality system procedures must exist and must be documented. Of course, the "range and detail of the procedures that form part of the quality system depend on the complexity of the work, the methods used, and the skills and training needed by personnel involved in carrying out the activity."[10] The control of these procedures is consistent with the assumed definition of quality system, where documented procedures is one of the stated components of a quality system.

Quality Plans

Paragraph 4.2.3 of ISO 9001 states that an organization "shall define and document how the requirements for quality will be met."[11] This statement forms the basis of quality planning. Quality planning should be done as soon as possible in the development of a new process. Of course, if processes are already in operation prior to seeking ISO 9001, documents are prepared for an existing system.

This planning activity could be considered the preparation of control plans or process quality plans which document how a process is controlled in order to achieve quality requirements. In either case, this documentation may or may not be in the Quality Manual depending on the size of the organization and the number of processes involved. If these quality plans are not found in the Quality Manual, then they would be in documents that are referenced by the Quality Manual. That set of documents then becomes a middle level of documentation between the top level (Quality Manual) and the bottom level (quality records).

From the design and engineering perspective, the quality plan would be associated with the procedures used by that department. That is, the quality plan applies to the operation and processes used in the design and engineering department. Of course, quality planning could also be performed for a specific development project. This is covered in a section of the ISO 9001 paragraph on design control, and this activity is called design and development planning.

In its consideration of the processes and procedures used in a department, ISO 9001 lists some factors that should be considered when doing quality planning. Some of these items listed in paragraph 4.2.3 are preparing quality plans, identifying processes and resources needed to achieve quality results, ensuring compatibility between various activities, identification of measurement requirements, identification of verification requirements, and identification and preparation of quality records.

For example, if one is creating a new group or team or process for performing design and engineering tasks, these aspects (at least) of the design process should be considered to determine if the new process is going to benefit the organization's achievement of quality. In order to demonstrate that this quality planning has been done, these considerations should be listed and evaluated by a document. The procedure to follow in creating this process could be in the Quality Manual or department procedures (Figure 10.1).

Contract Review

Paragraph 4.3 covers contract review. The contract is the document which specifies the agreement between a customer and a supplier. The contract is to define what the customer expects from the supplier. ISO 9001 is meant to apply to the supplier of the product to that customer; therefore, the contract review is from the supplier's perspective. That is, the organization that supplies the product (including the design and engineering of that product) is required to show that the contract has been reviewed prior to acceptance of the work.

For custom designed or heavily engineered products, the contract review is essential. Since the product is likely to be state-of-the-art or

---

QUALITY PLANNING

Consider the following sample issues at the beginning of a development project:

Processes:

Are any new procedures or processes required?

If yes,

Who will develop the procedures?
How will they be verified?
How will results be measured?
Who will be responsible for reporting progress?

Resources:

Are any new resources required?

If yes,

What new personnel are required?
What new hardware is required?
How will new hardware be acquired?
How will new hardware be calibrated and validated?
Who will be responsible for acquiring new resources?

Records:

What quality records will be created?
Who will be responsible for control of the records?
How will revisions be handled?

Interfaces:

Is a concurrent engineering approach to be used?
How will the project team communicate?

Figure 10.1   Quality planning.

unique in some way, it is vital that the design and engineering function in the supplier organization be represented in the contract review. The purpose of the contract review is to be certain that customer requirements can be met, and that the specification of customer requirements leads to the correct specification of the design requirements (in technical terms, for instance). Therefore, it is essential that the customer's expectations are clearly understood by the design and engineering function of the supplier's organization.

Since the contract review is a requirement in ISO 9001, the contract review must be documented. First of all, this means that the procedure (and what departments are to be involved) is to be documented. This contract review procedure is also to be controlled, and it should be easily incorporated into the Quality Manual. Secondly, the contract review must be documented in terms of its results. The contract review procedure should define how the results are documented. The results are considered a quality record (records generated by the quality system procedures); this record could be minutes of a contract review meeting, an approval form circulated through proper department managers, or memoranda indicating the results of a contract review.

ISO 9001 paragraph 4.3.2 lists some requirements for the contract review procedure. The purpose of this procedure is to ensure that the customer's requirements are adequately defined, that differences between the contract and any previous offerings are resolved, and that the supplier (or design and engineering organization) is capable of meeting the contract or accepted order requirements.

Design Control

Paragraph 4.4 covers design control. Obviously this is an essential section of the standard for designers and engineers and their management. In general, this section applies the basic ISO 9000 philosophy to the design and development process, that is, processes and procedures are to documented, controlled, and maintained.

However, in addition to controlling the process used for design and development, there are added requirements such as verification and validation activities related to the design itself (as opposed to just the process). The management aspects of design control are covered in subparagraphs on design and development planning, organizational and technical interfaces, design input, design output, design review, and design changes. Some of the design and engineering aspects of design control can be found in paragraphs on design review, design verification, and design validation (note that design review is included in both the designing and engineering aspect of design control as well as the management aspect of design control).

General

Paragraph 4.4.1 begins the presentation of the ISO 9001 design control requirements. This paragraph states that the organization "shall establish and maintain documented procedures to control and verify the design of the product."[12] This paragraph does not require that any certain method of design be used. By demanding that the method be documented, the organization is assumed to follow a consistent method and therefore be capable of consistently meeting customer requirements.

This is clearly another application of the basic ISO 9000 philosophy of documentation and control of processes that affect quality.

Design and development planning

Paragraph 4.4.2 demonstrates the quality planning concept applied to design. As mentioned earlier, an organization is expected to prepare quality plans for processes. In the case of design and development, an organization is expected to prepare design plans. According to this paragraph, the organization "shall prepare plans for each design and development activity. The plans shall describe or reference these activities, and define responsibility for their implementation."[13] Note that this paragraph repeats the need for assigning responsibility for tasks (similar to paragraph 4.1.2.1). Furthermore, this paragraph repeats the need for adequate resources (similar to paragraph 4.1.2.2); it states "design and development activities shall be assigned to qualified personnel equipped with adequate resources."

For a design and engineering department that consistently performs the same tasks for the same type of products, the design plan may just be the department's procedures. However, for a design and engineering department that performs different kinds of new product development, a design plan for each new product should be prepared (in addition to department procedures). This new product design plan should consider many of the same activities listed in paragraph 4.2.3 under quality planning. Some of these activities would include considering whether the new product requires any new controls, processes, resources, skills or equipment, considering whether the new product requires any new quality control, inspection, or testing equipment or techniques, and considering whether any new documentation, manuals, operator training, etc. is going to be required by the product.

In each of these planning activities, the intent is to assure quality by meeting customer's requirements. If an organization uses a meeting to consider these aspects of design planning, a quality record such as meeting minutes must be created and controlled. However, it would be best to create a standard procedure for creating a standard design plan document for each new product developed by the design and engineering department. Figure 10.2 presents a sample checklist for design and development planning. Such lists can be valuable in demonstrating that a consideration has been undertaken for this activity.

Organizational and technical interfaces

Paragraph 4.4.3 adds a special consideration for management in the design and engineering organization. This paragraph is titled "organi-

DESIGN & DEVELOPMENT PLANNING

Create a checklist for review when starting a new product design:

Processes:

☐ New Product Data Standards
New Engineering Standards
New Engineering Test Procedures
New Material Processing Procedures

Resources:

☐ New Test Equipment (hardware)
New Test Equipment (software)
Personnel Requisition
New Tooling
New Facilities

Records:

☐ New Quality Records
New Calibration Records
New Environmental Bulletins
New Manuals

Interfaces:

☐ New Product Design Team
New Concurrent Engineering Team
Newsletter
New Home Page

Figure 10.2   Design and development planning checklist.

zational and technical interfaces," and it acknowledges the cross-functional aspects of design activities. In the creation of a product, the expertise of marketing personnel, designers, engineers, technicians, and even scientists may be required. Therefore, it is important to emphasize proper interface and communications between these parts of the organization. According to this paragraph, "organizational technical interfaces between different groups which input into the design process shall be defined and the necessary information documented."[14]

The organizational interface can be defined by the organizational structure (or organizational charts) in the Quality Manual. The documented flow of information can be controlled by the use of memoranda or other internal correspondences. In some cases, a special format for documented information may be desirable. For instance, a development laboratory may have a test report format, or a materials laboratory may have a metallurgical report. If so, then these formats and reports must be given a standard procedure in the Quality Manual or another set of controlled documentation.

Design input

Paragraph 4.4.4 identifies an essential checkpoint in the design process. This checkpoint is called design input. Although a design and engineering organization may use many kinds of design processes, there must be a point in the process where the specifications of the design are established. These specifications must be documented and controlled. Thus the design input stage produces design input documentation. As stated in the standard, "design input requirements relating to the product, including applicable statutory and regulatory requirements, shall be identified, documented, and their selection reviewed by the supplier for adequacy."[15] Thus, the standard states that the establishment of the design input is to be reviewed, and this review should be used as an opportunity to verify that laws and regulations which may affect the design have been considered. Indeed, statutory and regulatory requirements are to be documented in the design input documentation.

There may be a number of mechanisms for documenting design input. One possibility is to use a contract. If the design and engineering organization processes a legal contract for each project, then a section of this contract may list the product specifications and design input. In fact, a section of paragraph 4.4.4 reminds the organization that "design input shall take into consideration the results of any contract review activities."[16] Another mechanism may be an internal document listing the design specifications and regulatory requirements. This internal document could then be reviewed and approved by interested parties (or even the external customer). Regardless of the mechanism used, the format of the design input and its associated documentation are to be included in the quality system documentation such as department procedures or the Quality Manual.

Of course, there could be changes to the design input. There could be changes due to the unavailability of resources, the design requirements may actually exceed the state of the art, or the customer (i.e., external customer) may need to renegotiate the requirements. In any case, the use of controlled documentation of the design input should allow for revisions. A legal contract could be revised or amended, or the internal design input documentation could be reissued and reapproved.

The use of care must be emphasized in establishing the design input. The design input becomes the starting point of the design process, and the design process is the starting point of satisfying the customer's needs and requirements. Therefore, changes to the design input after design work has already begun could cause an entire organization's process for customer satisfaction to be halted, and the costs associated with this starting over can be enormous. Although one cannot always prevent this from occurring, every possible opportunity should be taken to verify that all groups or departments or individuals related to the design understand the design input.

Design output

Paragraph 4.4.4 identifies another essential checkpoint in the design process; this stage produces documented output, so this checkpoint can be called design output. Note that the entire process between design input and design output is not mentioned in ISO 9001. Clearly, the process actually used to design a product is left to the organization. This is essential since the design process is going to differ depending on the organization, product, market, etc.

As with the design input, what is required is that the output from the design process be documented and that it be verified and validated. According to paragraph 4.4.5, "design output shall be documented and expressed in terms that can be verified against design-input requirements and validated."[17] Thus, although the use of specific techniques in design are not required by the standard, some techniques lend themselves to documentation and verification while other techniques may not.

As with design input, the mechanism for design output documentation is going to vary. In the case of a highly engineered new product, there may be various studies and reports concerned with the feasibility or performance of the new product. In most cases, release or package materials such as drawings, parts lists, material lists, etc. are going to be part of the design output.

The practice of reviewing and checking drawings or other documentation is a requirement of ISO 9001 if they are not reviewed as part of the design review process (as described later). This is a requirement since the last sentence of paragraph 4.4.5 states that "Design-output documents shall be reviewed before release."[18] Once drawings and documentation are released, design changes must be approved as well as reviewed since paragraph 4.4.9 states that "All design changes and

modifications shall be...reviewed, and approved by authorized personnel before their implementation."[19] Also, paragraph 4.5.2 states that
"documents and data shall be reviewed and approved for adequacy by
authorized personnel prior to issue."[20]

Regardless of the specific design output format, the standard states
the following about the design output:

> design output shall
>
> a) meet the design-input requirements;
> b) contain or make reference to acceptance criteria;
> c) identify those characteristics of the design that are crucial to the safe
> and proper functioning of the product[21]

In view of the listed demands related to design output, some special
design completion documentation needs to be considered. This documentation lists each of the design input items and then lists the completed design results (for instance, a design specification followed by a
design value calculated for the product). Furthermore, this list presents an opportunity to show the design results with acceptance criteria. These are criteria from design input based on customer requirements, regulatory requirements, or internal engineering standards.
Finally, this design completion documentation can specifically cover
safety critical items.

The creation of the design completion documentation could mark
the end of the design process. However, the processes of design
review, design verification, and design validation (which follow design
output) may require design changes. In this case, the design process
would be reentered and new design completion documentation would
have to be created. Design verification and validation activities are
listed in paragraphs 4.4.7 and 4.4.8.

Design review

Paragraph 4.4.6 specifies that the design process must be reviewed.
This is to be accomplished by holding meetings called design reviews.
The design review offers the first feedback opportunity to check that
the design is progressing toward having the design output meet the
requirements of the design input.

Although paragraph 4.4.7 specifies that a design review be held as
a check after the design output stage is reached, paragraph 4.4.6
offers some flexibility in holding design reviews. According to this
paragraph, "At appropriate stages of design, formal documented
reviews of the design results shall be planned and conducted."[22] Thus,
a design review can be conducted at various times. Often, a large

design project would have design milestones (corresponding to subassemblies or components or test program completions, for instance). A design review could then be held at the completion of these milestones. However, in smaller projects, a single design review meeting may be sufficient.

As paragraph 4.4.6 continues, there is an emphasis on the design review as an opportunity for bringing diverse expertise to bear. This paragraph states that "Participants at each design review shall include representatives of all functions concerned with the design stage being reviewed, as well as other specialist personnel."[23] The goal of the design review should be to make certain that the design is progressing toward the goal of meeting the design input requirements.

As always, the meeting of a quality system demand such as the design review must be documented. First, the procedure and timing for the design review must be documented as part of the quality system (either the Quality Manual or department procedures). Second, the results of the meeting (such as minutes) must be documented., maintained, and controlled to form evidence that the design review is being held. As paragraph 4.4.6 states, "Records of such reviews shall be maintained."[24] This documentation is considered a quality record since it is the documented result of carrying out a documented procedure.

Design verification and validation

The next two paragraphs of ISO 9001 specify checking activities for the design process. These activities are design verification and design validation; paragraph 4.4.7 covers design verification and paragraph 4.4.8 covers design validation. These two activities are probably the most important specific aspects of ISO 9001 for designers and engineers and their organization. Unfortunately, verification and validation have similar meanings and might be used interchangeably. However, ISO 9001 appears to make a clear distinction between these activities.

ISO 9001 makes a distinction between design verification and validation in terms of scope and methods. In terms of scope, design verification is supposed to verify that the design output meets the design input requirements. That is, the design that has been documented (design output) must be checked to see that it meets the design specifications (design input). As stated in paragraph 4.4.7, "design verification shall be performed to ensure that the design-stage output meets the design-stage input requirements."[25] Thus, the scope of the verification is entirely within the design process or the design and engineering department.

In terms of methods, paragraph 4.4.7 mentions such activities as design review meetings (as discussed earlier in this chapter), alternative calculations, and undertaking tests. Again, note that these activities are principally located within the responsibile design and engineering organization, and that their scope lies within the design input requirements.

The design validation activities, on the other hand, have broader scope and methods. As stated in paragraph 4.4.8, "Design validation shall be performed to ensure that product conforms to defined user needs and/or requirements."[26] Thus, design validation is like an "outer loop" checking activity; while design verification is like an "inner loop." Since the customer's or user's needs are translated into the design input, and the design output, in turn, is to satisfy the design input, one would think that the design output would necessarily meet customer's requirements. However, one purpose of design validation is to ensure that the translation from customer input to design input has been appropriately successful.

Since design validation is the outer loop, it would necessarily follow design verification. This is confirmed in the notes for paragraph 4.4.8; here it is stated that "Design validation follows successful design verification."[27] Also, since the design validation is for checking the product that has been designed against end user conditions, these notes also state that "Validation is normally performed on the final product," and it stated that "Validation is normally performed under defined operating conditions"[28] (such as those specified in a contract or customer specifications).

Design verification

Assuming that the nature of design verification in ISO 9001 is understood, a more in-depth analysis is possible. Design verification can be considered the basic checking activity of designers and engineers. It is hoped that most design and engineering organizations already utilize design verification activities regardless of the introduction of the ISO 9001 standard.

One specific method of design verification is required by ISO 9001. This is the use of a design review meeting (as defined earlier). The purpose of such a design review meeting is to check that the design output meets the design input (as opposed to other design reviews that may verify other aspects of design). This design review would investigate the drawings, calculations, test data, field data, etc. associated with the completion of the design stage and compare them with the design specifications and acceptance criteria. This design review meeting is an excellent opportunity to consider or brainstorm issues such as design manufacturability, safety, reliability, cost, failure modes and effects, etc. (although optimally these issues have been considered throughout the design process).

Recall that paragraph 4.4.5 specified that design output identify those characteristics of the design that are essential to safe use of the design; these acceptance criteria certainly must be checked against the design as completed. Finally, the design review must be documented (using meeting reports or minutes) to create a quality record that can be used to demonstrate that this design verification activity has been completed.

In addition to the design review meeting described above, other forms of design verification are probably necessary. Activities such as reviewing, checking, or approving design calculations can be considered design verification if the design calculation output shows the specific design parameters to be satisfied (i.e., acceptance criteria). Without the acceptance criteria along with the design calculations, design approval can only show that another person or party has reviewed the results. It does not necessarily show that the design output has met the design input.

Although designers may use "checked by" and "approved by" procedures, engineering calculations (analyses considered to be beyond the scope of designers) should consider various other forms of verification.. In some organizations, designers may be using methods that are actually developed by engineers. Therefore, verifying the design calculations requires that engineering methods or algorithms be verified as well. The need for this deeper level of verification is indicated by paragraph 4.4.7, which states that design verification may include "performing alternative calculations."[29]

Therefore, when new design methods such as formulas and computer programs are developed, these methods should be verified for correct results. In terms of formulas, their development can be documented and compared to other published methods for calculating the same parameters (certainly a wealth of different published methods and techniques exist for most areas of engineering science). In terms of computer programs for design and engineering, a whole new area of quality management is needed to make sure that the latest version of verified (and validated) software is being used. The reader is directed to the Bibliography for references on this topic.

Another method of design verification is testing. As opposed to testing a production model in the customer's intended use environment (i.e., design validation), design verification testing would be used to verify engineering methods or to provide proof of concepts and prototypes. For instance, testing might be used to measure stresses in a part after a new formula or algorithm is developed. Assuming these results are documented and the results are within acceptable limits, the new formulas can be shown to be verified based on the tests. This testing could be performed addition to the "alternative calculation" method mentioned earlier.

Design verification testing could also be used directly to check that design output meets design input requirements. In this case, the specific parameters or specifications of design input would be measured. Subsequently, the results could be compared to the acceptance criteria for each of the parameters. The comparison of the results and acceptance criteria can then be documented by a controlled form or report. This documentation would form a quality record to demonstrate design verification. Of course, this method could be used for some parameters while other forms of design verification could be used for other parameters; the design review would still be needed.

Any discussion of alternative calculations and testing introduces the problem of "engineering judgment" and statistical techniques. For instance, if a new method for calculating a design parameter is created, and the results of the method are compared to previously published methods, how does one decide if the new method is better or worse than the previous method? In terms of testing, how does one assess the accuracy of an engineering method versus the test data? Certainly no set of measurements of complex characteristics (such as stresses, loads, pressures, temperatures, etc.) is going to be exactly the same as the engineering prediction technique.

It should be the responsibility of the design and engineering organization to make decisions concerning the use of various engineering methods. Certainly, the ISO 9001 registrar (or third-party auditor) is not going to provide any guidance in deeming a specific engineering technique good or bad. The only concern of the registrar is that whatever judgment is made must be documented and considered as part of the design review process. Therefore, the discussion of whether a design is acceptable (in any design verification activity) should still allow for the "engineering judgment" aspects of the design and its calculations.

Of course, the use of statistical analysis should be considered in this context. Testing results should always include an uncertainty analysis. Furthermore, appropriate statistical techniques should also be applied so that those reviewing the design are aware of the accuracy of the test data. Indeed, designers and engineers should consider studying the statistical methods that have been refined and propagated recently in the quality control field for understanding the capability and validity of production and control systems.

There are still other forms of design verification noted in ISO 9001. According to paragraph 4.4.7, design verification may include "comparing the new design with a similar proven design, if available."[30] This form of design verification could be based on a design review meeting or a written report which compares and contrasts a given new design to previous designs. This form of design verification could also be used to exempt portions of a design from more rigorous analysis and verification, since some portions of the design may be similar or identical to existing, proven products.

As always, if this comparison technique of design verification is used, there must be a documented record of the comparison. In such documentation, the design being reviewed and the design forming the basis of comparison would have to be clearly identified. Furthermore, the excerpt from paragraph 4.4.7 states that the previous design be similar and proven. Thus, the documentation must indicate how the previous design has been considered proven. This proof could be documentation of the operational record or the success of testing, calculations, and correlations of the previous design.

In summary, design verification should be a primary focus of designers and engineers in an organization considering or implementing the ISO 9001 standard for their quality system. Design verification is a basic checking and approving activity for designers and engineers. This activity should seek to install confidence in the work of designers and engineers within the organization, its customers, and an auditing registrar. The results of these activities must be documented, maintained, and controlled in accordance with procedures covering quality records.

Design validation

As presented earlier, design validation follows successful design verification. According to paragraph 4.4.8, "Design validation shall be performed to ensure that product conforms to defined user needs and/or requirements."[31] It is assumed that the design would now be in a production form with all design changes for satisfying design verification completed.

Perhaps the best form of design validation would be to test a sample product from a production run in the user's defined environment. However, this may not always be possible. If the product cannot be tested, an approval procedure or a review meeting could be used. The design output stage documentation would be compared to the customer's requirements (as defined from the contract review, for instance). The design validation review and approval procedure would not compare with the design input or specifications, however, since that check is supposed to be accomplished by design verification.

The goal of the design validation procedure is to ensure that the product is going to satisfy the customer, not just the design specifications. Notes from paragraph 4.4.8 indicate that the design validation "is normally performed on the final product, but may be necessary in earlier stages prior to product completion."[32] In this case, during the design process itself, the design could be evaluated in the scope of customer requirements (going beyond the assumed scope of design verification). The design validation procedure would also be performed after the design process and design verification are completed. Paragraph 4.4.8 also indicates that "multiple validations may be performed if there are different intended uses."[33]

The design validation procedure should also consider (again) areas such as safety, storage, service, manufacturability, etc. since it is likely to be easier to change the design at this stage as opposed to later stages. Finally, whatever procedure is developed for design validation, it must be a documented procedure in the Quality Manual or the department procedures.

Design changes

The final paragraph of ISO 9001 under Design Control is design changes (4.4.9). This paragraph states that "design changes and modifications shall be identified, documented, reviewed and approved by authorized personnel before their implementation."[34] Normally this type of control is going to already be available to the design and engineering organization. Certainly, indicating the revision level on drawings, along with revision checking and approval boxes on drawings, is common practice. However, in implementing or studying ISO 9001, a design and engineering organization should consider whether the process of requesting changes to drawings is actually understood and properly documented.

As usual, the process for requesting changes to a design should be documented. This design change or engineering change procedure could be a department procedure, but since various departments (such as manufacturing and marketing) may also need to request design changes, this procedure may be best placed in the Quality Manual. The design change procedure should use a form or other media so that a quality record is created. This form can then be cataloged and thus meet the demand for "documentation" of design changes. Furthermore, the catalog can use an identification scheme (numbering, codes, etc.) so that the demand for identification of design changes can be met.

Note that the design change procedure should allow for rejection of the request. That is, not all requests for changes to a design may be granted. Such a request should meet the demand of reviewing design changes. Since this decision is clearly going to impact the quality of the product or design, the personnel responsible for this decision should be identified in the Quality Manual or engineering department procedures. If a design change request is rejected, the form or other quality record should be signed and a notification and reasons for rejection sent to the request initiator. This action could complete the design change review process or the requestor could decide to bring the matter to a higher management level (in accordance with paragraph 4.1.2.1 on Responsibility and Authority).

Finally, the design change procedure must provide approval by authorized personnel prior to implementation of the change. If a change is approved, the form or quality record can be signed and a design change process can begin. Depending on the size and complexity of the organizational structure, there may be a process for design or engineering change request, a process for engineering change review, a process for engineering change revalidation, and a process for engineering change notification. A simplified design change process is charted in Figure 10.3.

The changes in the design documentation (drawings, parts lists, bills of materials, etc.) would then have their own checking and approval process after the design changes are made prior to release. As mentioned earlier, in order to meet the demand for having authorized personnel provide approval, these personnel should be identified in the engineering department procedures or Quality Manual.

Document and Data Control

ISO 9001 makes some specific demands on documentation and data used throughout the organization. These demands are presented in paragraph 4.5 which goes beyond the scope of design control (paragraph 4.4). These demands apply to the entire organization or company. Indeed, paragraph 4.5.1 states that "supplier shall establish and maintain documented procedures to control all documents and data that relate to the requirements of this American National Standard including, to the extent applicable, documents of external origin such as standards and customer drawings."[35] Clearly the designers and engineers in an organization or company are going to deal regularly with this type of documentation and data. Since this statement says that all documents and data that relate to the standard are to be controlled by documented procedures, and since ISO 9001 applies to the design and engineering area uniquely, it may be assumed that all the

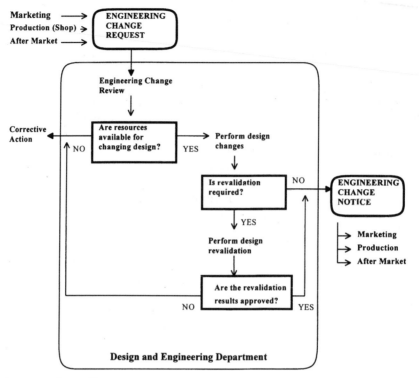

Figure 10.3   Design Change sample process.

documents and data in the design and engineering area need to have controlling procedures.

Thus, all the information in the design and engineering area should have a known status or disposition. Although this seems to be an excessively broad demand, one could simply implement a policy that those documents in the design and engineering area which are not clearly marked by a CONTROLLED DOCUMENT designation are FOR REFERENCE ONLY. If there are archive files or libraries of information in the design and engineering area that are not marked FOR REFERENCE ONLY, signs should be placed on the archive files or library areas clearly indicating that the entire contents of the file or library is under the FOR REFERENCE ONLY designation.

Paragraphs 4.5.2 and 4.5.3 of ISO 9001 cover more demands of the control of documents and data. For designers and engineers, most of these needs are already covered by demands of design control (design input, design output, design changes, etc. as already presented). For example, paragraph 4.4.5 on Design Output states that "output documents shall be reviewed before release."[36] Also, paragraph 4.4.9 on

Design Changes states that changes shall be "documented, reviewed, and approved by authorized personnel before their implementation."[37] These concepts seem to already cover the requirements of later paragraphs 4.5.2 and 4.5.3 when they state that "documents and data shall be reviewed and approved for adequacy by authorized personnel prior to issue"[38] and "Changes to documents and data shall be reviewed and approved by the same functions/organizations that performed the original review and approval, unless specifically designated otherwise."[39]

Although designers and engineers should simply be aware of and abide by procedures created for the control of documents and data, engineering management should also carefully review the rest of the material in paragraph 4.5.3. Management should consider such topics as printed versus electronic media, record retention, and treatment of obsolete documents and data. The primary goal of this consideration should be to see that the correct and latest version of all documents and data is easily available to all members of the design and engineering organization. Of course, this is a worthy goal irrespective of ISO 9001, and resources should be allocated as needed to strive for achievement of this goal.

Inspection and Testing

The next area of the ISO 9001 standard that may be of interest to design and engineers is paragraph 4.10 on inspection and testing. The information in this paragraph may be of interest since tests may be specified as part of the design process. Recall that tests and demonstrations could be used for design verification, and that design validation may also involve testing. Design verification testing would be analogous to prototype testing, and this activity may involve a designer or engineer. In terms of inspection, a production run sample of a new product design may be inspected or audited to verify that the design as documented in the design output is actually what is found from the production output.

Another possible scenario for designers or engineers is the request for laboratory-type testing of a new process, material, or design method. In this case, the designer or engineer needs to consider standard or controlled procedures for specifying the test, planning for it, and documenting its results.

As a general rule, any testing connected with the design of the product should be performed according to controlled procedures. According to paragraph 4.10.1 of ISO 9001, "[the] supplier shall establish and maintain documented procedures for inspection and testing activities in order to verify that the specified requirements for the product are met."[40] Testing and inspection to support design activities should be considered requirements for the product.

The design and engineering organization has already been shown to require documented procedures for its functions. Therefore, if testing is performed within the design and engineering organization [such as a research and development (R&D) shop or development laboratory or testing area] these areas should not be exempted from the development and maintenance of documented procedures.

The procedures developed in this area should include a formal request for testing. Such a testing request should include the requirements and acceptance criteria for the testing, reference written procedures for the actual testing performed, and require a formal output or report format for presenting the results of the testing (including the uncertainty of results). This formal output can then be referenced by the design output from designers and engineers during product development, and the ISO 9001 requirement for controlled testing should be met.

After specifying that inspection and testing be performed in a controlled manner, paragraphs 4.10.2–4.10.5 present a variety of requirements for the inspection and testing of incoming, in-process, and completed work. These paragraphs should be reviewed by engineering or laboratory management to be certain that the requirements are being met in the context of the design and engineering organization's product and activities. Many of these paragraphs probably apply to the testing of actual products and materials as opposed to developmental testing in the prototype or product development stage. However, these paragraphs should still be reviewed carefully to be certain all requirements are being met.

Paragraph 4.11 of ISO 9001 should also be reviewed by engineering or laboratory management. This paragraph covers control of test and inspection equipment. These requirements are centered on calibration activities and assuring that results obtained from inspection and testing can be relied upon. As stated in this paragraph, "[the] supplier shall establish and maintain documented procedures to control, calibrate, and maintain inspection, measuring, and test equipment (including test software) used by the supplier."[41] In addition to initial calibration of this equipment, paragraph 4.11.1 indicates that the equipment (and software) shall be rechecked at prescribed intervals.

Paragraph 4.11.2, which also deals with testing, lists further considerations pertaining to inspection and testing activities. These considerations include the following items that should be reviewed by engineering management for relevance with respect to design testing activities:

The supplier shall:

a) determine the measurements to be made and the accuracy required...;

b) identify all inspection, measuring, and test equipment that can affect product quality; and calibrate and adjust them at prescribed intervals,...;

c) define the process employed for the calibration of inspection, measuring, and test equipment,...;

d) identify inspection, measuring, and test equipment...;

e) maintain calibration records...;

f) assess and document the validity of previous inspection and test results when inspection, measuring, and test equipment is found to be out of calibration;

g) ensure the environmental conditions are suitable...;

h) ensure that the handling, preservation, and storage of inspection, measuring, and test equipment is such that the accuracy and fitness for use are maintained;

i) safeguard inspection, measuring, and test facilities, including both test hardware and software, from adjustments that would invalidate the calibration setting.[42]

In addition to reviewing these requirements in the context of inspection and testing related to product development, engineering management should note that ISO 9001 specifically mentions test software. One can assume that this reference is meant to apply to the data acquisition type of computer software which is so common in testing activities. This software presents a number of unique problems, including the need to periodically revalidate the software. This reference to test software can also be interpreted to mean that, as test devices or instruments need to be recalibrated according to a prescribed schedule, so the data acquisition software needs to be compared with benchmark results on a prescribed schedule.

Conclusion

This chapter has presented some basic information specifically highlighted from the ISO 9001 standard with respect to designers and engineers. Most of these topics are interpreted with respect to the concept of quality system. The development of sample procedures and documents based on ISO 9001 are found in Chapters 3 through 7.

Finally, Figure 10.4 presents a list of the basic concepts in this chapter. Figure A.1 in Appendix A also includes a listing of the ISO

Highlighted ISO 9001 Requirements for Designers and Engineers	
Requirement	Summary (with ANSI/ASQC Q9001-1994 paragraph)
Quality Manual	The top level of documented procedures (4.2.1).
Quality Policy	A basic statement of policy to comply with ISO 9001 which is found in the Quality Manual (4.1.1).
Organizational Structure	A set of organizational charts and departmental definitions found in the Quality Manual (4.1.2.1).
Management Representative	A person primarily responsible for the quality system and its continuous improvement (4.1.2.3).
Quality Assurance Function	A resource or group defined in the Quality Manual for at least auditing and tracking the quality system (4.1.2.2).
Management Review	A procedure defined in the Quality Manual for periodic critical review of the quality system; must be recorded (4.1.3).
Contract Review	A procedure defined in the Quality Manual for critical review of a contract; must be recorded (4.3).
Quality Plans	A procedure defined in the Quality Manual for preparing plans related to new processes; must be recorded (4.2.3).
Design Plans	An engineering department procedure for preparing plans related to new products or new designs (4.4.2).
Design Input	An engineering department procedure that defines a design "spec" document (4.4.4).
Design Review	An engineering department procedure for critical review of a given design; must be recorded (4.4.6).
Design Verification	An engineering department procedure for checking that design and engineering work meets "specs" (4.4.7).
Design Validation	An engineering department procedure for checking that new product meets customer requirements (4.4.8).
Design Changes	An engineering department procedure for controlling and documenting revisions to product designs (4.4.9).
Inspection and Testing	Procedures for controlling and documenting design and other testing including calibration and revalidation (4.10).

Figure 10.4   ISO 9001 Requirements review.

9001 sections by paragraph and the sample procedures and documents found in Chapters 3 through 7.

References

1. ASQC, *ISO 9001 (ANSI/ASQC Q9001-1994) Quality Systems—Model for Quality Assurance in Design, Development, Production, Installation, and Servicing*, ASQC, Milwaukee, Wisconsin, 1994, paragraph 1, p. 1.
2. ASQC, ISO 9001 paragraph 4.1.1, p. 1.
3. ASQC, ISO 9001 paragraph 4.1.1, p. 1.
4. ASQC, ISO 9001 paragraph 4.1.2.1, p. 1.
5. ASQC, ISO 9001 paragraph 4.1.2.1, p. 2.
6. ASQC, ISO 9001 paragraph 4.1.2.3, p. 2.
7. ASQC, ISO 9001 paragraph 4.1.3, p. 2.
8. ASQC, ISO 9001 paragraph 4.1.2.2, p. 2.
9. ASQC, ISO 9001 paragraph 4.2.1, p. 2.
10. ASQC, ISO 9001 paragraph 4.2.2, p. 2.
11. ASQC, ISO 9001 paragraph 4.2.3, p. 2.
12. ASQC, ISO 9001 paragraph 4.4.1, p. 3.
13. ASQC, ISO 9001 paragraph 4.4.2, p. 3.
14. ASQC, ISO 9001 paragraph 4.4.3, p. 3.
15. ASQC, ISO 9001 paragraph 4.4.4, p. 3.
16. ASQC, ISO 9001 paragraph 4.4.4, p. 3.
17. ASQC, ISO 9001 paragraph 4.4.5, p. 3.
18. ASQC, ISO 9001 paragraph 4.4.5, p. 3.
19. ASQC, ISO 9001 paragraph 4.4.9, p. 4.
20. ASQC, ISO 9001 paragraph 4.5.2, p. 4.
21. ASQC, ISO 9001 paragraph 4.4.5, p. 3.
22. ASQC, ISO 9001 paragraph 4.4.6, p. 4.
23. ASQC, ISO 9001 paragraph 4.4.6, p. 4.
24. ASQC, ISO 9001 paragraph 4.4.6, p. 4.
25. ASQC, ISO 9001 paragraph 4.4.7, p. 4.
26. ASQC, ISO 9001 paragraph 4.4.8, p. 4.
27. ASQC, ISO 9001 paragraph 4.4.8, p. 4.
28. ASQC, ISO 9001 paragraph 4.4.8, p. 4.
29. ASQC, ISO 9001 paragraph 4.4.7, p. 4.
30. ASQC, ISO 9001 paragraph 4.4.7, p. 4.
31. ASQC, ISO 9001 paragraph 4.4.8, p. 4.
32. ASQC, ISO 9001 paragraph 4.4.8, p. 4.
33. ASQC, ISO 9001 paragraph 4.4.8, p. 4.
34. ASQC, ISO 9001 paragraph 4.4.9, p. 4.
35. ASQC, ISO 9001 paragraph 4.5.1, p. 4.
36. ASQC, ISO 9001 paragraph 4.4.5, p. 3.
37. ASQC, ISO 9001 paragraph 4.4.9, p. 4.
38. ASQC, ISO 9001 paragraph 4.5.2, p. 4.
39. ASQC, ISO 9001 paragraph 4.5.3, p. 4.
40. ASQC, ISO 9001 paragraph 4.10.1, p. 6.
41. ASQC, ISO 9001 paragraph 4.11.1, p. 6.
42. ASQC, ISO 9001 paragraph 4.11.2, p. 7.

Helpful References
to ISO 9001

This appendix presents some helpful references with respect to ISO 9001. These references are intended to help designers and engineers easily find sections of the ISO 9001 standard that are of the most interest.

First, a number of figures within the various chapters of this book list ISO 9001 paragraphs of most interest to designers and engineers. These figures are Figure 3.2 (showing the paragraphs that lead to the ideal ISO 9001 development process), Figure 3.9 (showing the organizational demands aspect of ISO 9001), and Figure 5.4 (showing the ISO 9001 design verification needs).

Second, Figure A.1 presents a table of the sample documents in this book along with the corresponding figure numbers. The figure numbers are given in sequential order by chapter. The paragraph or subparagraph within ISO 9001 which the sample document is attempting to address is then listed.

SAMPLE DOCUMENT	FIGURE (in this work)	ISO 9001 Reference and Paragraph (ANSI/ASQC Q9001-1994)
Process Review Checklist	3-10	4.4.2 Design and development planning
Design Specification Master Document	4-1	4.4.4 Design input
Design Specification	4-2	4.4.4 Design input
Development Project Master Log	4-4	4.5.2 Document and data approval and issue
Engineering Standard	5-3	4.4.1 General (DESIGN CONTROL)
Verification Study	5-5	4.4.7 Design verification
Historical Verification Form	5-6	4.4.7 Design verification
Engineering Test Procedure	5-7	4.10.1 General (INSPECTION AND TESTING)
Engineering Test Request	5-10	4.10.1 General (INSPECTION AND TESTING)
Engineering Test Request Log	5-11	4.12 INSPECTION AND TEST STATUS
Engineering Test Report	5-12	4.10.5 Inspection and test records

Figure A.1   Sample documentation in ISO 9001.

SAMPLE DOCUMENT	FIGURE (in this work)	ISO 9001 Reference and Paragraph (ANSI/ASQC Q9001-1994)
Prototype Verification Study	5-13	4.4.7 Design verification
Design Review Notice	5-16	4.4.6 Design review
Design Stage Design Review	5-17	4.4.7 Design verification
Product Data Standard	6-3	4.4.5 Design output
Released Drawing Format	6-4	4.4.5 Design output
Design Validation Form	7-1	4.4.8 Design validation

Figure A.1 Sample documentation in ISO 9001 (cont'd).

Engineering in ISO 9001

What Is the Difference Between Design and Engineering?

Considering that this is a work on ISO 9001 for designers and engineers, it may be of some benefit to consider the differences between designers and engineers. Some aspects of ISO 9001 apply to these technical professionals in different ways. Often the engineer in an organization is more responsible for the development of methods used to design, whereas the designer is actually responsible for determining the detailed configuration of a given product.

In some organizations these tasks may be combined into a single person or department. Also, in recent years designers and engineers, as well as technical support and manufacturing support personnel, have been reorganized into product development teams (as opposed to being constantly separated by technical expertise). This procedure is often referred to as *concurrent engineering*.

In any case, the discussions presented in this book assume that the reader is familiar with some distinction between designers and engineers. One could consider that this distinction may not have been difficult in past generations, but the advent of computer technology has blurred the distinction between these disciplines and helped to remove their distinction from the rest of organizations. The designer may now use computer software to perform complex engineering calculations (which previously would only have been done by engineers). On the other hand, the engineer may now use computer software to do detail design work with preprogrammed design experience and computer-aided drafting (which previously would only have been done by designers).

Regardless of the danger of generalizations, for the purpose of this book, one may consider an engineer to be someone whose basic task is to apply mathematics and sciences to the development of new analytical techniques for product design. Designers, in turn, use those analytical techniques, other existing techniques, and past experience to create documentation and graphics which describes new products. This is a gross simplification, but may prove helpful to those without experience in the area of new product development. Another generalization is that the engineer may be viewed as being more analytical, while the designer may be more practically oriented.

This simplified view of engineers and designers can be helpful in viewing the management model presented in ISO 9001 for design and development. There is a requirement for design verification in the context of design calculations (a more engineer-oriented activity), and there is a requirement for design validation in the context of design performance (perhaps a more designer-oriented activity).

However, one must note that there is no specific mention of engineering as a separate activity from design in ISO 9001. This is fortunate since the processes of engineering and design are becoming more "concurrent" (in the sense of concurrent engineering as discussed earlier) and there may be a wide difference in the interpretation of engineering versus design.

With respect to design and engineering, ISO 9001 is basically interested in the processes which produce new products that are intended to satisfy defined requirements. Whether engineers or designers should perform particular aspects of this process is not specified by ISO 9001 (as long as the engineers and designers are considered qualified and trained).

What Is Development?

Another problem presented by a book such as this is that one is forced to consider a definition of design and development. Since ISO 9001 is supposed to be widely applied to all kinds of activities (including nonindustrial activities), it may be unfair to limit design and development to physical products. However, one may assume that a book intended for designers and engineers (such as this one) is going to be speaking to an industrial product development audience.

It is assumed that in the reader's organization there is a physical product to be designed by creating drawings, models, and other documentation. Furthermore, this product is going to be built, tested, and perfected by the same organization. The new product documentation would then be used by a manufacturing facility to produce the product and then release it to a customer. In the age of concurrent engineering, computer simulations, and virtual corporations, this can be a limiting definition; however, it is a reasonable assumption that any industrial audience will be familiar with the process of design and development as presented.

We should also note that designers and engineers are to be considered part of a larger process of product development and introduction. Although determining a product's basic configuration is the primary focus for designers and engineers, the larger process would typically include marketing, finance, manufacturing, and product or field support.

One of the most important larger scope activities with respect to engineering and design is marketing. As noted in previous chapters, achieving quality by meeting customers' requirements takes for granted that the customer's requirements are documented and understood. This is a basic task of the marketing function. The end product or output from the marketing function should be a documented new product proposal or contract of some kind.

The output from marketing naturally becomes the input to the design and development process performed by designers and engineers. However, note that this distinction between engineering and marketing is also becoming less defined. Again, in no small part because of the power and availability of computer software, it is possible to consider design and engineering calculations during the traditional marketing stage. Therefore, designers and engineers may also be involved in the process of establishing customer requirements, new product proposals, or contract details. There should be no difficulty with this development with respect to ISO 9001 as long as the integrated organizational structure is defined and documented, and the various personnel are qualified and trained adequately.

Development Project Scenario

For the benefit of those not familiar with product design activities, consider a generalized example of the product design and development process from the industrial perspective. The first step would be the decision to attempt to create a new product. Note that this new product is rarely going to be revolutionary. More likely, this new product is going to be an extrapolation of past market and technological experience. Also note that a decision has only been made to attempt to create the product. In other words, the new product may or may not be feasible; the market for the product may or may not exist.

Since the marketability and decision to proceed with development of the product is going to depend on the price of the product, it may be necessary to consider some initial designing of the product in order to estimate the price. This phase may be considered the market research or feasibility study. This phase is going to be generally driven by the marketing function or department within the organization, but optimally preliminary design work performed within marketing function can be transferred to the design and engineering function if a decision is made to proceed with the development project.

If the market research has found that the product is feasible and a market may exist, the marketing function (with the input of various areas of the organization) decides that the design and development process should be begun for the proposed product. The marketing function along with the design and engineering function then creates a design specification or "spec." This specification provides the information about what the product must do in order for the product to meet the goals of the market research (or to satisfy the external customer). The design specification is a document that provides the approval for the larger expenditure of design and engineering activities.

The design and development process now begins in earnest. Using the design specification as the guideline, a design can be created. Note that the new product may demand more or less of engineering versus design input. If the organization's product is highly engineered, then a significant amount of analytical calculations are going to be needed. For example, in the author's experience in the gas compression equipment sector, modeling of gases at high pressures is required for each and every quotation and order. In this case, there would be a high level of involvement by engineers in modeling these gas flows for each design. If the organization's product is not that highly engineered, engineers may only be needed occasionally when standard design techniques fail to produce adequate results or when unexplained failures occur during design and development.

Different products may also require large amounts of simulation. This simulation attempts to predict the behavior of the product. Simulation can be based on computer software or it may be based on laboratory models which are often then analyzed by computers. Simulation is another example of a task in the design and development process for the highly engineered product.

Thus, for the highly engineered product, there is a whole phase of design and development devoted to complex calculations and fundamental investigations. The output from this phase may only be a preliminary design. During this phase, engineers may create sketches (manual or electronic) using the basic outline and configuration of the product. The designers then take the basic outline and create all the necessary details to complete the product design.

At some point, the task of creating the physical, manufacturable product must be performed. The plan is that designers, in close collaboration with engineers, create drawings or three-dimensional models of a prototype product. The prototype is the output from a preliminary design phase. The prototype may be studied in a laboratory environment or it may be evaluated in an actual market environment (i.e., test marketing). In most cases, the prototype needs to be revised in light of actual performance results.

Optimally, at some point it is decided that the prototype product is acceptable. It is determined that the actual physical product can succeed in the marketplace. The product is considered able to meet the customer requirements laid out in the design specification. This process can be considered the final design phase. Note that some products (particularly those that are highly engineered) may not be mass produced. In this case, it may be known at the end of the preliminary design phase if customer requirements will be met, since the customer may witness or evaluate the actual results of a prototype product. In the case of the mass-produced consumer product, this is not the case. Instead, a marketing decision may be needed to proceed with the manufacture and sale of the product at the completion of a design phase.

Assuming that it is going to be introduced to the market, the product is then going to be manufactured. The product is going to go through a manufacturing development phase. In this phase, the design and development phase comes to a close. Hopefully manufacturing expertise has been an integral part of the design and development phase so that the best methods of manufacture can be determined and implemented quickly.

If the product enjoys some measure of success in the marketplace, it is going to enter service. Although the work of designers and engineers is generally over, there may still be activities requiring their attention. In some cases, the product may need refinement. In this case, engineers and designers would need to evaluate the ramifications of these revisions. Engineers and designers may also need to be involved with litigation related to the product. Although these activities impact designers and engineers, they are not directly part of an organization's product design and development processes. Instead, the impact of design refinement and litigation is going to be a consideration throughout all the processes (market research, preliminary design, final design, and product support).

Engineering Quality Assurance

The last view of engineering in ISO 9001 concerns the vital importance of quality in the process of design and engineering. Whether the analytical methods of engineers or the practical experience of designers are being considered, it is essential not to leave the achievement of quality to the quality assurance function within an organization.

Although design and engineering involves art as well as science, and these processes tend to be approached with an attitude of mystery, there is no reason that designers and engineers cannot be held responsible for quality concepts such as customer satisfaction and process improvement. Every designer and engineer has "internal customers" (those who receive and use their work within the organization). Just as the organization seeks to meet the expectations of the external customer for the benefit of the entire organization, designers and engineers must seek out, define, refine, and whenever possible document the expectations of their internal customers.

A quality assurance process such as knowing customer requirements may seem like overhead to designers and engineers, but their first-hand involvement with product design tends to lead them to focus to a great extent on product design, possibly to the exclusion of issues related to customer satisfaction and process improvement. Such a focus is often necessary to achieve the needed concentration on highly technical tasks. However, designers' and engineers' energy can easily be wasted if their concentrated effort is spent pursuing a poorly understood or misunderstood directive from the designer's or engineer's internal customer. Therefore, they should be encouraged to focus on their customer as well as product design.

Another engineering quality assurance process (besides knowing one's internal customer) is continuous improvement. Again, although design and engineering may seem mysterious, there is no reason that they cannot be understood and then improved by a given organization's management. Improving these processes can mean making flowcharts to study internal processes and producing and distributing surveys to internal customers. It can even involve measuring, displaying, and attempting to improve measured values of performance (also called metrics). Some examples of metrics might be time to market, number of engineering change requests (due to engineering error) made per month, or monetary value of component failures during design and testing phases. This process improvement concept may seem like overhead, but it may also provide a competitive advantage if applied conscientiously.

Conclusion

This appendix has presented some background information with respect to design and engineering in ISO 9001. This information is meant to assist the reader in understanding the processes of product design and development that relate to the ISO 9001 standard. This type of information tends to be assumed by the presentation throughout this book. This appendix presents the generalization that engineers tend to be responsible for analytical product design techniques, while designers tend to be responsible for product design details. This appendix also presents the idea that quality assurance concepts such as internal customer requirements and process improvement can be applied to design and engineering processes.

Bibliography

American Nuclear Society, ANSI/ANS-10.4-1987, *Guidelines for the Verification and Validation of Scientific and Engineering Computer Programs for the Nuclear Industry,* ANS, La Grange Park, Illinois, 1987.

American Society of Mechanical Engineers, ASME NQA-2a-1990, *Part 2.7 Quality Assurance Requirements of Computer Software for Nuclear Facility Applications,* ASME, New York, 1990.

American Society for Quality Control, ANSI/ASQC Q9000-1-1994, *Quality Management and Quality Assurance Standards—Guidelines for Selection and Use,* ASQC, Milwaukee, Wisconsin, 1994.

American Society for Quality Control, ANSI/ASQC Q9004-1-1994, *Quality Management and Quality System Elements—Guidelines,* ASQC, Milwaukee, Wisconsin, 1994.

American Society for Quality Control, ANSI/ASQC Q9003-1994, *Quality Systems— Model for Quality Assurance in Final Inspection and Test,* ASQC, Milwaukee, Wisconsin, 1994.

American Society for Quality Control, ANSI/ASQC Q9001-1994, *Quality Systems— Model for Quality Assurance in Design, Development, Production, Installation, and Servicing,* ASQC, Milwaukee, Wisconsin, 1994.

American Society for Quality Control, ANSI/ASQC Q9002-1994, *Quality Systems— Model for Quality Assurance in Production, Installation, and Servicing,* ASQC, Milwaukee, Wisconsin, 1994.

Basili, V. R., "Models and Metrics for Software Management and Engineering," *ASME Advances in Computer Technology,* Vol. 1, pp. 278–289, 1980.

Belie, R. G., "Computers in American Industry—The Race with Confusion," *ASME Computers in Engineering 1987,* Vol. 2, pp. 137–140, 1987.

British Standards Institution, BS 7165: 1991, *Recommendations for Achievement of Quality in Software,* BSI, Linford Wood, Milton Keynes, England, 1991.

Durand, I. G., et al., "Updating the ISO 9000 Quality Standards: Responding to Marketplace Needs," *Quality Progress,* July 1993, pp. 23–28.

Gerhart, S. L., "Fundamental Concepts of Program Verification," *ASME Advances in Computer Technology 1980,* Vol. 1, pp. 305–314, 1980.

Holman, J. P., *Experimental Methods for Engineers,* 3rd ed., McGraw-Hill, New York, 1978.

International Organization for Standardization, ISO 9000-3:1991(E), *Quality Management and Quality Assurance Standards—Part 3: Guidelines for the Application of ISO 9001 to the Development, Supply, and Maintenance of Software,* ISO, Geneva, Switzerland, 1991.

Powers, J., "TQM in Software Development Organizations," *Quality Progress,* July 1993, pp. 79–80.

Richter, H. P., "Effective Computer Program Development and Use," *ASME Advances in Computer Technology 1980,* Vol. 1, pp. 335–340, 1980.

Ridlon, S., "Qualification and Training of Users of Finite Element Programs," *ASME Advances in Computer Technology 1980,* Vol. 2, pp. 227–229, 1980.

Schoonmaker, S. J., "Engineering Software Quality Management," *ASME Petroleum*

Division, Vol. 43 (Computer Applications and Design Abstractions 1992), pp. 55–63, 1992.

Schoonmaker, S. J., "Engineering Software Quality Management and ISO 9000," *ASME CIE Division—Engineering Data Managment Proceedings 1993*, pp. 195–204, 1993.

Schoonmaker, S. J., "Techniques in Engineering Software Quality Management," in Leondes, C. T. (ed.), *Control and Dynamic Systems*, Vol. 60-0, San Diego, Academic Press, 1994, pp. 289–328.

Schuster, D. J., "Introduction to Computer Standards Use," *ASME Advances in Computer Technology 1980*, Vol. 2, pp. 418–422, 1980.

Smith, G. E., "The Dangers of CAD," *Mechanical Engineering*, February 1986, pp. 58–64.

Index

ABOUT THE AUTHOR

Stephen Schoonmaker is currently Manager of Engineering Systems at Grove Worldwide, a gobal supplier of mobile hydraulic cranes and aerial work platforms. This assignment has blended his background in engineering, software development, and advanced technology implementation. Prior to this assignment, Schoonmaker was a Senior Engineer at the Engine Process Compressor Division of Dresser-Rand. In that position, he wrote the analytical engineering software and created engineering standards for the quality management of this class of software. The latter activity gave Schoonmaker his experience with quality assurance, management, and the ISO 9000 standards.

Schoonmaker has been active in the American Society of Mechanical Engineers since 1982. After receiving a degree in mechanical engineering from WPI in Worcester, Massachusetts, in 1984, he received the Williston Medal from the ASME. He is a past president of the New York Southern Tier Section of ASME, and he is currently an alternate delegate to its National Agenda Committe. While at Dresser-Rand, Schoonmaker was involved with ASME Pressure Vessel Code calculations for reciprocating compressor pulsation dampener design. In 1994, Schoonmaker created and conducted a number of ASME professional development short courses on ISO 9001 and engineering software documentation. In addition to his work with ASME, Schoonmaker is a member of the American Society for Quality Control.